Let the Evidence Speak

Alan Jessop

Let the Evidence Speak

Using Bayesian Thinking in Law, Medicine, Ecology and Other Areas

 Springer

Alan Jessop
Durham University Business School
Durham University
Durham, United Kingdom

ISBN 978-3-319-71391-5   ISBN 978-3-319-71392-2   (eBook)
https://doi.org/10.1007/978-3-319-71392-2

Library of Congress Control Number: 2017960423

Printed on acid-free paper

This Springer imprint is published by the registered company Springer International Publishing AG part of Springer Nature.
The registered company address is: Gewerbestrasse 11, 6330 Cham, Switzerland

Contents

1   Introduction   1

Part I   Likelihood

2   Whose Car?   7

3   Bayes' Rule   15

4   Track Record   23

5   Game Show   31

6   Margin of Error   39

7   Five Thoughts About Likelihoods   53

Part II   Base Rate

8   Diagnosis   69

9   Information   83

10   Being Careful About Base Rates   97

11   Independence   111

12   Review   127

Part III   Application

13   Who Wrote That?   139

14   Wood, Trees and Wildlife   155

15   Radiocarbon Dating   173

16 Bayes and the Law   189

17 The Selection Task   207

18 Conclusion   223

A Formula for Bayes' Rule   225

Chapter 1
Introduction

Bayes is all the rage!

You'll read this, or something like it, from time to time. This example is from the opening sentence of a book by Luc Bovens and Stephan Hartmann [1]. In full the sentence reads "Bayes is all the rage in philosophy". You may not be much interested in philosophy, but that's just one of the places where Bayes' Rule makes an appearance. There are others, not only in books written for academics.

For instance, Angela Saini wrote a piece in *The Guardian* about the use of Bayes' Rule in a court of law under the headline *The formula for justice*. She writes that Bayes' Rule was

> Invented by an 18th-century English mathematician, Thomas Bayes, this calculates the odds of one event happening given the odds of other related events [2].

And here's Tim Harford writing in a piece called *How to make good guesses* that when we have two pieces of information

> Logically, one should combine the two pieces of information ... There is a mathematical rule for doing this perfectly (it's called Bayes' rule) [3].

Finally, describing the application of Big Data to online advertising, Cathy O'Neil writes

> The data scientists start off with a Bayesian approach, which in statistics is pretty close to plain vanilla. The point of Bayesian analysis is to rank the variables with the most impact on the desired outcome. Search advertising, TV, billboards, and other promotions would each be measured as a function of their effectiveness per dollar. Each develops a different probability, which is expressed as a value, or a weight [4].

All of which is fine, but just what is Bayes' Rule?

It's been with us for about 300 years and every so often enjoys some popular prominence. Articles and books such as those above are not written for a technical audience. They may have whetted your appetite—sounds good but how does it work? You want to know more but you don't want a statistics text. This book is for you.

$$\sim \bullet \sim$$

© Springer International Publishing AG, part of Springer Nature 2018
A. Jessop, *Let the Evidence Speak*, https://doi.org/10.1007/978-3-319-71392-2_1

Decisions should be made based on evidence.

A doctor may use a test such as a blood test or MRI scan when diagnosing what illness a patient might have. Experimental results show the diagnostic power of the test, its ability to provide evidence for the doctor who then must decide what to tell the patient. Some tests are very informative, some less so.

Police investigating a crime and juries trying an accused have to use evidence too. This is of variable quality, ranging from DNA matching (pretty good) to eyewitness identification (not so good), and yet decisions of guilt and innocence are made.

What is common is that someone—a doctor, a juror—makes a decision based on what they believe to be true and this belief is based, in part at least, on evidence. Bayes' Rule provides the necessary link between the evidence and what to believe based on that evidence. The better the evidence the stronger is the belief that the patient is ill or the accused is guilty. The strength of the evidence, and so of our belief, is described using probability.

Bayes' Rule takes us that far. What to do, what action to take, is your decision. You will have a measure of your justified belief to help you.

~ • ~

There are many fine books and research papers you could read about Bayes' Rule and its applications but they tend to be written in the formal language of mathematics. They may assume that you know more than you do. We all hit a barrier at some point.

The motivation for this book is that it is possible and useful to describe for any reader, whatever their (which is to say, your) background, what Bayes' Rule is and how it works. Bayes' Rule is usually presented as a formula. For many this is not an encouragement to read further. But a great many real applications, not just text book illustrations, need no more than simple arithmetic. To show the calculations I have used a table—the Bayes Grid. All you need to know is how to add, multiply and divide. Think of a very simple spreadsheet.

Bayes' Rule is also a way of thinking, of forcing you to answer questions

What do I know about this problem?
What evidence do I have and how good is it?
What alternative explanations might account for the evidence?

Asking the right questions is always a good idea, even if you sometimes need help with the maths.

~ • ~

This book is organised in three parts. The first sets out the basic relation between evidence and the alternative accounts for that evidence. The second extends the basic framework to bring in other information you may have. This may be based on data or on judgement. You need to know how to deal with the issues raised by this useful extension.

Up to this point the calculations have been easy. The emphasis has been on thinking about how to use evidence, asking the right questions. Many applications are not so straightforward either because the maths is more difficult or because the decision problem is, by its nature, more complex. The chapters in this third part cover both.

In summary, the three parts are

Likelihood   The key to it all. The necessary description of the relation between the evidence you have, what might account for it, and what you are justified in believing.

Base rate   Base rates enable you to use contextual or judgemental information as well as evidence. Bayes' Rule becomes a means of learning, modifying your starting belief in light of new evidence. Bringing in judgement is quite natural, but care is needed. We may not be as good as we think at expressing our judgement in the form of probabilities.

Application   Many of the applications of Bayes' Rule use mathematical models which are daunting for non-mathematicians. However, the underlying principles are the same as in the earlier parts. These final chapters show how some Bayes' thinking helps structure problems and some of the modelling issues which follow.

    The closing two chapters, on law and the psychology of reasoning, do not involve heavy maths. They further emphasise the benefits of using Bayes' Rule as a way of thinking.

Notes and references are there to provide hints and reading should you wish to go further. Some of the references may not be for you but none are needed for you to read this book and, I hope, benefit from it.

~~~ • ~~~

References

1. Bovens L, Hartmann S (2003) Bayesian epistemology. Clarendon, Oxford, p v
2. Saini A (2011) The formula for justice. Guardian Magazine, 2 Oct 2011, p 12
3. Harford T (2016) How to make good guesses. Financial Times Weekend Magazine, 27/28 Feb, p 45
4. O'Neill C (2016) Weapons of math destruction: how big data increases inequality and threatens democracy. Allen Lane, London, p 74

Part I
Likelihood

Chapter 2
Whose Car?

Evidence provided to the police is often imperfect. Even witnesses who are quite sure of what they saw or heard may be wrong. How should their evidence be evaluated?

Let us suppose.

There has been a robbery at a jewellers in Stockholm. Inspector Larsson and his squad have narrowed the suspects down to just two, Jan and Stig. Larsson is sure that one of them is guilty but there is no conclusive evidence to decide between them. Circumstantial evidence puts them both near the crime scene at about the right time, but that's not enough. Then a witness, Ingrid, comes forward. She was walking in the street where the robbery occurred and saw a car driving very fast away from the jewellers. She only got a very brief look at the car and all she could say was that it was blue. Stig's car is blue and Jan's car is green. But thanks to the creative efforts of car manufacturers it's not quite as clear cut as that. A list of car paint codes shows Albi Blue, Storm Blue, Mountain Blue, Odyssey Blue, Spray Blue. . .. There are twenty-six shades with blue in the name, and then there's Aqua and others. There are only eleven which are green, including Nordic Green. Larsson smiles. Probably all that can be said is that Stig's car is more blue-ish and Jan's car more green-ish.

So, how can Ingrid's evidence be evaluated?

Colour is defined by wavelength measured in nanometres (nm). One nanometre is a billionth of a metre. The visible spectrum is from about 400nm to about 700nm. Shorter wavelengths, less than 400nm, are called gamma rays and X rays. Longer wavelengths are used for microwave and radio. In the visible part of the spectrum ranges of wavelength are given names which are the colours with which we are all familiar (think rainbow)[1]:

---

[1]These values are from the NASA website. https://science-edu.larc.nasa.gov/EDDOCS/Wavelengths_for_Colors.html. Accessed 14 September 2017.

A. Jessop, *Let the Evidence Speak*, https://doi.org/10.1007/978-3-319-71392-2_2

violet	about 400
indigo	about 445
blue	about 475
green	about 510
yellow	about 570
orange	about 590
red	about 650

The way we map this infinite variety into a small and manageable number of colours is called *categorical perception*. The same idea can be applied to sounds and other stimuli. There is inevitably some uncertainty, some vagueness, about where to draw the line. We may be inconsistent, applying different labels to the same stimulus on different occasions. Different people will apply different labels. When Ingrid said "blue" what did she mean?

While browsing the web Larsson's trusted colleague, Mankell, comes across a study [1] which may help. In it a number of people were shown colour cards for a short time and then asked to say whether what they saw was blue or green. The results are shown as *identification functions*. For any colour shown by its wavelength on the horizontal axis we can see the percentage of times it was identified either as "blue" or as "green".

Figure 2.1 shows the function for the label "blue". As the wavelength increases fewer people say "blue". For a colour with wavelength 491nm expect that fifty percent of people would call it "blue".

Fig. 2.1 Identification function for "blue"[2]

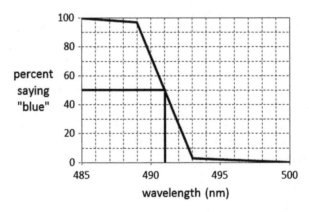

[2]This and other graphs in this chapter use information from Fig. 2a of Bornstein and Korda [1]: 212.

Figure 2.2 shows the identification functions for both "blue" and "green". The two lines are complementary, mirror images. For any wavelength the two percentages sum to one hundred since only those two alternative names—"blue" and "green"—were used in the experiment.

Fig. 2.2 Identification functions for "blue" and "green"

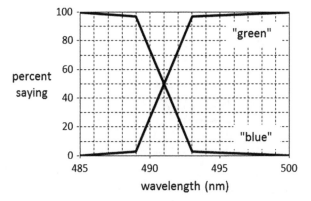

This graph might be helpful in assessing just what weight should be given to Ingrid's evidence. The actual colours of the two cars are known, the wavelengths have been measured by the crime scene investigation team, and so the percentage of times that each car would be expected to be called "blue" or "green" can be read from the graph.

There is no reason to think that Ingrid differs in any relevant way from the subjects of the original experiment. This graph will do as a basis for assessing how useful her evidence is in discriminating between Stig's car and Jan's car.

We can mark the actual colour of Stig's car (490nm). Our best estimate is that seeing Stig's car seventy-four percent of people will call it "blue" and twenty-six percent will say "green" (Fig. 2.3).

Fig. 2.3 Stig's car

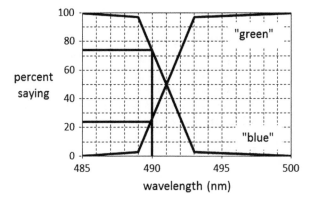

For Jan's car (colour is 492.5nm) Fig. 2.4 shows that only sixteen percent of people would call it "blue" and eighty-four percent "green".

Fig. 2.4 Jan's car

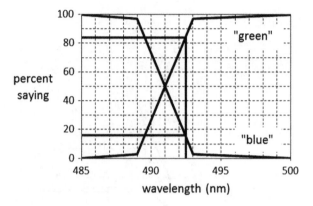

Because they assume that Ingrid is typical of the subjects tested in the experiment it seems clear to Larsson and Mankell that these results may be used to describe the accuracy of her assessment: when she says "blue" the odds that the car is Stig's car rather than Jan's car are 74:16. To put it another way, the chance that the car was Stig's car is eighty-two percent.[3] The chance that it was Jan's car is eighteen percent. Larsson and Mankell conclude that based on Ingrid's evidence the probability that it was Stig's car that she saw is about eighty percent. Not bad.

~ • ~

The two policemen did what seemed obvious. Were they right?

Yes, they were.

Although they didn't know it, they had used Bayes' Rule. It was the intuitively obvious thing to do.

~ • ~

This is fine so far as it goes. We have a good idea of the strength of this evidence but how should it be used together with what else we know, or suspect? How to combine different sources of evidence?

The natural language of the courtroom is not numeric and neither judges nor jury members (nor many of us) are used to this form of reasoning. But courts do hear evidence from expert witnesses, much of which is based on quantitative analysis, and juries do make decisions.

---

[3]Odds and probabilities both express uncertainty. The conversion from one to the other is easy. $74/(74 + 16) = 82\%$ and $16/(74 + 16) = 18\%$.

The usefulness and, which is different, the admissibility of arguments about the possible statistical basis of evidence and the justifiable evaluation of that evidence is subject to quite a bit of argument. Justification is important.

~ • ~

Stig, Jan, Ingrid and the rest are fiction. But the possible use of Bayes' Rule to assess evidence and, more problematically, to present the analysis in court, are not.

In 1996, Denis Adams was accused of rape. The victim said her attacker was in his twenties. Adams was 37. The victim was unable to pick Adams in an identity parade. Adams' alibi was that he had spent the night in question with his girlfriend. But DNA evidence at the scene was a good match with Adams' DNA, though there was some dispute about the correct match probability. The match probability is the probability that someone picked at random would match the DNA profile found at the crime scene. The prosecutor said 1 in 200 million but the defence argued for a much lower figure of 1 in 20 million or perhaps even 1 in 2 million. All other evidence supported Adams' claim of innocence. What was the jury to make of all this? Professor Peter Donnelly of Oxford University was permitted by the court to show the jury how Bayes' Rule could be used to evaluate the evidence presented to them. Adams was convicted. He appealed against his conviction.

The Court of Appeal noted that at the original trial no advice was given to the jury about what they should do if they chose not to use Bayes' Rule. A retrial was ordered.

At the retrial the court asked that Professor Donelly and other statisticians on the prosecution's side cooperate to provide a guide to help the jury. Although Donnelly had reservations a method was agreed. Jury members were required to complete questionnaires to illustrate how Bayes' Rule could be used (Professor Donelly gives a brief account of all this in the Royal Statistical Society's journal *Significance* [2]).

Adams was convicted again. He appealed again. The appeal was not upheld. But the Court of Appeal made observations critical of the use of Bayes' Rule. For example

Jurors evaluate evidence and reach a conclusion not by means of a formula, mathematical or otherwise, but by the joint application of their individual common sense and knowledge of the world to the evidence before them.

Individual jurors might differ greatly not only according to how cogent they found a particular piece of evidence (which would be a matter for discussion and debate between the jury as a whole), but also on the question of what percentage figure for probability should be placed on that evidence.

Different jurors might well wish to select different numerical figures even when they were broadly agreed on the weight of the evidence in question.

The general drift was that the use of Bayes' Rule, or any mathematical model, would illegitimately undermine "an area peculiarly and exclusively within the province of the jury, namely the way in which they evaluate the relationship between one piece of evidence and another". In other words, the mental processes of twelve members of the public can be trusted to make some quite complicated judgements. That is sometimes inevitable, of course, but where a little calculation might help it seems harsh to deny the jurors the use of it.

The court decided against the use of Bayes' Rule in assessing DNA evidence. But, even if jurors had some appreciation of probabilities of events familiar to them or which they might easily imagine—tossing a coin, choosing a card—they (and we) are unlikely to have an appreciation of what a probability of 1 in 200 million, or 0.000000005, means. In the trial this very low rate implied that the number of people in the UK matching Adam's DNA profile was small, just one or two. They might be children or older folk and so not suspects. Or they might be related to Adams. He did, in fact, have a half-brother whose DNA profile was not known.

If the match rate was as high as 1 in 2 million then there might be about thirty or so people who would match: a different picture.

Guidance along these lines has subsequently been issued to judges so that they might help jurors, and themselves, get a grip on these very low probabilities.

~ • ~

This isn't a problem which will go away. In 2010 a convicted killer, "T", took his case to the Court of Appeal. One of the pieces of evidence against him involved his Nike trainers and the likelihood that shoe prints at the scene of the crime matched his shoes. The judge believed that the expert witness had incorrectly calculated the match probability. The case was quashed.

Professor Norman Fenton of Queen Mary, University of London, pointed out that the conclusion was not well justified and, surprise, proposed that Bayes' Rule would help.[4] He, with others, built a model based on the rule to help in judicial decision making. Their model allows the combination of evidence which is based on data and evidence which relies on judgement.[5]

We'll return to this problem of using Bayes thinking in court in Chap. 16.

~ • ~

The same ideas should help doctors to make diagnoses. You can easily imagine other applications.

And just think where all that big data analysis might lead. Simply having more data is not enough, we still need to have a way of reaching justifiable decisions.

It should be clear that numerical analyses do not replace judgement. But where the application of a little mathematics can help that judgement it would be wilful to reject it out of hand. There is pretty good evidence that we human beings aren't too smart when it comes to dealing with numerical information unaided by some calculation. Just what to do with the answers is where judgement comes to the fore.

~ • ~

---

[4]Some of the issues raised are discussed in [3].

[5]More details from their company, Agena. Bayesian Network and Simulation Software for Risk Analysis and Decision Support. http://www.agenarisk.com/. Accessed 14 September 2017.

The concerns of the courtroom were a long way from the mind of the Reverend Thomas Bayes when he devised his rule for reasoning from evidence to belief. But just what is Bayes' Rule? And who was Bayes?

~~~ ••• ~~~

References

1. Bornstein MH, Korda NO (1984) Discrimination and matching within and between hues measured by reaction times: some implications for categorical perception and levels of information processing. Psychol Res 46(3):207–222
2. Donelly P (2005) Appealing statistics. Significance 2(1):46–48
3. Fenton NE, Neil M, Hsu A (2014) Calculating and understanding the value of any type of match evidence when there are potential testing errors. Artif Intell Law 22(1):1–28

Chapter 3
Bayes' Rule

In the previous chapter the two policemen had evidence that was not precise. The evidence they had was an eyewitness report that a car was "blue" and they wanted to see how strongly this pointed to Stig's car, which was blue, or to Jan's car, which was green. This is typical of many problems we face: we have to decide between a number of alternatives (which car?) based on some imprecise evidence ("blue"). To judge just how useful the evidence is we need to know the relation between each of the alternatives and the evidence. This goes by many names: accuracy, track record, diagnostic power and so on.

The idea is simple enough. If I don't know how much you know about the Nigerian economy why should I ask you about it? Or, if you give your opinion anyway, how seriously should I take what you say? I should find out about your track record. How good a pundit are you? Often we don't, of course. What do you know of your doctor's track record? Dealing with this sort of judgemental evidence raises all sorts of issues about credibility, bias and much else.

Fortunately, for a great many problems we have data on which to base an evaluation of the relation between what we have as evidence and the alternative causes or explanations of that evidence. Larsson and his team show how we might use evidence. It is a straightforward case which we can put into a more formal, though still simple, framework which will be useful for other problems.

Remember that the paint used by the makers of Stig's car would be called "blue" by seventy-four percent of people and "green" by the rest. Only sixteen percent of people would call Jan's car "blue". Table 3.1 shows this information.

© Springer International Publishing AG, part of Springer Nature 2018   15
A. Jessop, *Let the Evidence Speak*, https://doi.org/10.1007/978-3-319-71392-2_3

alternatives: car colour	evidence: witness statement		sum
	"blue"	"green"	
Stig's car	74	26	100%
Jan's car	16	84	100%

Table 3.1 Colour identification depends on car

Each row of the table shows how likely is the evidence—"blue" or "green"—depending on which car was seen. The sum of each row is one hundred percent. This is because only these two values of evidence are thought to be possible. We have ruled out the possibility that a witness will say that either car was red, for instance. So, a witness must say either that the car was blue or that it was green: there are no other alternatives, no other colours are considered. These two distinct values cover all possibilities. That is why the sum is one hundred percent.

This is a table presenting some experimental data.

The witness, Ingrid, was not part of the study described in the research but we have no reason to believe that she is different from the study participants in the way she identifies colours. It is reasonable to say that the likelihood, the probability, that she would have called Stig's car blue is seventy-four percent.

Just pause for a minute. The shift in perspective from data summary to prediction is common enough, but just check it. To take one of those simple examples that stats teachers use to illustrate a point, think of a deck of cards. There are thirteen cards in each of four suits. That is data summary. The probability that you will pick a diamond is a quarter, twenty-five percent. That is a probability assessment for the result of your decision: a prediction.

Just as a diamond is one of the cards in the deck so Ingrid is one of the population of Stockholm. But it's a bit more than that. We reasonably assume that one deck of cards is like any other. There is no reason to believe that the deck from which you choose is different from any other deck. In all cases the probability of picking a diamond is a quarter. And so we assume that the population from which Ingrid is drawn, the residents of Stockholm, is no better or worse at identifying colours than the population from which the people taking part in the experiment came. This is plausible.

~ • ~

Each row of the table is a *probability distribution*. The values in the row show how likely evidence is for each of the two alternatives, the two cars in this case. These probability distributions are called *likelihood distributions* and are the key to describing how evidence is related to alternative explanations of that evidence.

The table showed probabilities as percentages which sum to one hundred but they could also be shown as decimal fractions that sum to one, as in Table 3.2.

| alternatives: | evidence: witness statement | | sum |
car colour	"blue"	"green"	
Stig's car	0.74	0.26	1
Jan's car	0.16	0.84	1

Table 3.2 Likelihood distributions

Use either. Most people find percentages easier to read though sometimes decimal fractions are easier for calculations. However you show them likelihoods are the key to using evidence.

∼ • ∼

Of all the possible different values of evidence, "blue" and "green", we have one, "blue". What should we believe? Do as the police did: believe in the alternatives according to how likely each is to explain the evidence. It makes sense to express this belief as a probability distribution, so rescale the likelihood values to sum to one hundred percent. Figure 3.1 shows this simple rescaling.

Fig. 3.1 What to believe given Ingrid's evidence

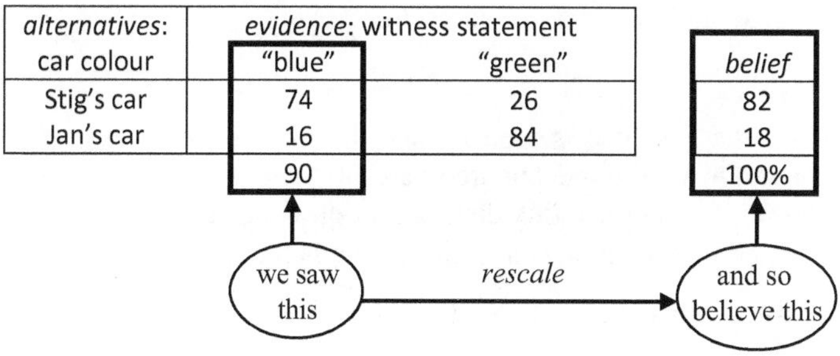

∼ • ∼

Using the word belief for these probabilities is important. It emphasises that what you believe is personal. It may be that where the evidence and likelihoods are uncontroversial we would all believe the same thing; that the probability of a coin coming down heads is fifty percent, for instance. But this is not always so. Different people may identify a different set of alternatives—Stig, Jan and Mats. Some of these differences can be accommodated in the model or extensions of it. The following chapters will show how.

∼ • ∼

We now have a framework for reasoning with evidence.

A simple table is all that we needed to give a structure to our problem.

The model was just "fill the rows then look at the columns". There will be a little more to it than that, but that is the essence.

Here are the components of the model as shown in the table.

1. *Each row shows an alternative*.
   What might explain the evidence; a particular deck of cards, a particular car?
   Have we missed any possible alternative explanations?

2. *Each element in a row shows a possible value of the evidence*.
   What evidence might we collect; a card, a colour?
   The list shows all possible values so that one, but only one, will be seen.

3. *Values in rows show likelihoods*.
   For each cell in a row write how likely it is that that evidence will be seen if that alternative is the true alternative explanation.
   Call these numbers, the entries in each row, the *likelihood* of each piece of evidence.
   Make sure each row sums to one hundred percent to give a complete likelihood distribution for each alternative.

4. *Rule*.
   Once the evidence is seen rescale the likelihoods in the corresponding column to sum to one hundred percent.
   The result is a probability distribution showing the degree of belief you are justified in having that each alternative is the true alternative.

The key word is **justified**.

~ • ~

To summarise, remember the very important rule which tells us how to make sense of what we see

> believe in alternative causes or explanations in proportion to the
> degree to which they explain the evidence

or   **belief is proportional to likelihood**

This, in one form or another, is Bayes' Rule, named for its proposer. But just who was Bayes?

~ • ~

If ever you are in London go to Bunhill Fields on the edge of the City near the cluster of tech start-ups known as silicon roundabout (British irony, perhaps). Here, just off the

City Road, you find a small burial ground in which rest the remains of nonconformist churchmen. It is now a public park. Although appropriately modest this cemetery is home (if that is the right word) to William Blake and John Bunyan and Daniel Defoe. Here also is the tomb of the Bayes and Cotton families. On the top of this tomb is an inscription commemorating its 1969 restoration made possible by "subscriptions from statisticians worldwide". This generosity was in recognition of the contribution (which we have just begun to examine) of one of the Bayes family, a priest.

Fig. 3.2 Thomas Bayes (1702–1761)[1]

Thomas Bayes (Fig. 3.2) was born in 1702. His father was Joshua Bayes, a nonconformist minister at Leather Lane, in Holborn, London. In 1731, following a private education, Thomas also was ordained as a Presbyterian minister and took up a ministry at Tunbridge Wells, in Kent. He had a lifelong interest in mathematics and statistics, and was elected a Fellow of the Royal Society in 1742. He retired from the ministry in 1752 and remained in Tunbridge Wells until his death in 1761.

He published just two papers during his lifetime: in 1731, *Divine Providence and Government Is the Happiness of His Creatures*, and, five years later, *An Introduction to the Doctrine of Fluxions, and a Defense of the Analyst*, this being an attack on Bishop Berkeley following the Bishop's attack on Newton's calculus.

After Bayes' death his friend Richard Price, also a Bunhill Fields resident, sent a further paper by Bayes to the Royal Society and in 1763 it was published in the *Philosophical Transactions of the Royal Society of London* as *Essay Towards Solving a Problem in the Doctrine of Chances*. It is this paper which contains the famous rule.[2]

~ • ~

---

[1]Original source unknown. Wikimedia Commons. https://commons.wikimedia.org/wiki/File: Thomas_Bayes.gif. Accessed 16 September 2017.

[2]There are a number of accounts of Bayes life and work. Here are two: Barnard [1] and Dale [2]. Another, which concentrates more on the post-war spread of Bayes' ideas, is Fienberg [3].

The rule was independently proposed some years later by the French mathematician Pierre-Simon Laplace (Fig. 3.3). Born of humble origins in 1749 at Beaumont-en-Auge in Normandy, Laplace's intellectual achievements were, to put it mildly, considerable, but he seems to be in other respects not possessed of wholly admirable qualities. His vanity and arrogance were matched by his ability to blow with the wind so as not to fall out with patrons and politicians. Given the bloody times through which he lived in revolutionary France, not least the time of the Reign of Terror in 1793, this may have been no bad thing. He wisely agreed with the scheme for the new revolutionary calendar even though he knew that the length of the proposed year was incompatible with astronomical observations then available.

Fig. 3.3   Pierre-Simon Laplace (1749–1827)[3]

It was, in fact, in the field of mathematical astronomy that Laplace made his most distinguished contribution, the five volume *Traité du Mécanique Céleste,* which, a little unusually for the time, contained no mention of God. When Laplace went to present this work to Napoleon the following well known exchange occurred:

NAPOLEON   Monsieur Laplace, they tell me you have written this large book on the
                         system of the universe, and have never even mentioned its Creator.

LAPLACE   Sir, I had no need of that hypothesis.

In 1812 Laplace published his *Théorie Analytique des Probabilités*. In the second volume (of only two this time) he set out his definition of probability.

---

[3]Image from Album du centenaire by Augustin Challamel (1889). Wikimedia Commons. https://commons.wikimedia.org/wiki/File:AduC_197_Laplace_(P.S.,_marquis_de,_1749-1827).JPG. Accessed 13 October 2017.

This was, it transpired, the same as Bayes' Rule, which was unknown to him. In fact, the rule was not even called Bayes' Rule at that time. It was so named a few years later by another mathematician, Poincaré, like Laplace a Frenchman, thereby avoiding yet another Anglo-French tiff.

~ • ~

Here is the problem with which Bayes was concerned, as he set it out:

> *Given* the number of times in which an unknown event has happened and failed: *Required* the chance that the probability of its happening in a single trial lies somewhere between any two degrees of probability that can be named.

We may say

> *Given* the number of people who say "blue" rather than "green": *Required* the chance that the probability that one person will say "blue" when seeing a colour of known wavelength lies somewhere in a specified range.

This problem is a little more subtle than the analysis of Ingrid's witness statement. We did not conclude that the probability that the car was Stig's car was between seventy percent and ninety percent but only that it is eighty-two percent. It is likely that what we would actually say is that is about eighty-two percent, the "about" signalling some caution.

We'll think about this later, but for now we'll stick with the simpler idea.

~ • ~

The rule that

belief is proportional to likelihood

is uncontroversial. It is the base for statistical inference, for deciding what we should believe based on available data. There is, of course, a huge literature on this and the necessary mathematics.[4] There are many excellent books you can read if you are interested. Here are just a few, roughly in order of mathematical explicitness.

McGrane [4] All narrative with no maths at all. A good read.

Phillips [5] An old book, so might be hard to find. I read and liked it when it was first published. Still do. About the level you might expect for an introductory stats class for those with not much maths.

---

[4]Most of the applications of Bayes' Rule are described in the books and journals particular to the application; forensic science, archaeology and so on. You might also find the website of the International Society for Bayesian Analysis (https://bayesian.org/) and their journal Bayesian Analysis useful, though most contributions are quite technical.

Stone [6] A short introduction. Quite friendly.

Lee [7] Another short introduction for those with a maths background.

Gill [8] Not a short introduction. Read this if you're happy with Lee and want more.

O'Hagan and West [9] For those already familiar with Bayesian analysis who want to see the variety of practical applications.

But this is not a book which requires you to have a background in mathematics or statistics; not at all. The emphasis is on using a simple framework to think more clearly and ask useful questions. This doesn't require you to do much more than the calculations in the table introduced in this chapter. We'll use this table a lot so give it the shorthand name Bayes Grid. It's simple and works well for a lot of problems.

In the next three chapters we will see how to use likelihoods based on a track record of past performance, some careful reasoning, and a statistical model.

~~~ ••• ~~~

References

1. Barnard GA (1958) Thomas Bayes—A biographical note. Biometrika 45(3&4):293–315
2. Dale AI (2003) Most honourable remembrance: the life and work of Thomas Bayes. Springer, New York
3. Fienberg SE (2006) When did Bayesian inference become "Bayesian"? Bayesian Anal 1(1):1–40
4. McGrane SB (2011) The theory that would not die: how Bayes' Rule cracked the Enigma code, hunted down Russian submarines, and emerged triumphant from two centuries of controversy. Yale University Press, New Haven
5. Phillips LD (1973) Bayesian statistics for social scientists. Nelson, London
6. Stone JV (2013) Bayes' Rule: a tutorial introduction to Bayesian analysis. Sebtel Press, Sheffield
7. Lee PM (2004) Bayesian statistics: an introduction, 3rd edn. Wiley, Chichester
8. Gill J (2015) Bayesian methods: a social and behavioral sciences approach, 3rd edn. Chapman and Hall/CRC, Boca Raton
9. O'Hagan A, West M (eds) (2010) The Oxford handbook of applied Bayesian analysis. Oxford University Press, Oxford

Chapter 4
Track Record

Experimental results are one way of finding the relation between what people say and alternative accounts of why they said it. Likelihoods quantify that relation and make the results usable. That's what the Stockholm police did to find how likely it was that Ingrid said "blue" or "green". The data showed what a number of people, the sample, said and how their judgements varied with the colour they saw. We don't always have a convenient experimental sample. Sometimes we have advice or prediction from just one source; a financial advisor, a doctor, the weather forecast or perhaps, if football is your thing, a football pundit.

~ • ~

Step forward Mark Lawrenson, Lawro to his many fans. When Liverpool signed him from Brighton and Hove Albion for £900,000 in 1981 he was the most expensive defender in Britain. Following a distinguished playing career for Liverpool and for the Republic of Ireland he is now one of the BBC team of football experts.

As well as providing his thoughts on the matches he also makes predictions. Each week he predicts the results of the ten Premiership games to be played the following weekend. He makes his predictions on Thursday to go on the BBC football website on Friday, so he has no information of last minute team changes.

This is a bit of fun and not Lawro's main job but it seems reasonable to believe he does his best. So, how good are Lawro's predictions? Table 4.1 shows his predictions and the actual scores for the ten matches on the first day of the 2013–2014 season.

© Springer International Publishing AG, part of Springer Nature 2018
A. Jessop, *Let the Evidence Speak*, https://doi.org/10.1007/978-3-319-71392-2_4

			Lawro	Score	
Liverpool	v	Stoke City	1-0	1-0	✓✓
Arsenal	v	Aston Villa	2-0	1-3	✗
Norwich	v	Everton	2-1	2-2	✗
Sunderland	v	Fulham	1-1	0-1	✗
West Bromwich Albion	v	Southampton	2-1	0-1	✗
West Ham	v	Cardiff City	2-0	2-0	✓✓
Swansea City	v	Manchester United	1-1	1-4	✗
Crystal Palace	v	Tottenham Hotspur	1-2	0-1	✓
Chelsea	v	Hull City	2-0	2-0	✓✓
Manchester City	v	Newcastle United	2-0	4-0	✓

Table 4.1 Predictions and results for the first week

The predictions were perfectly accurate (✓✓) in three cases. In another two the result was correctly predicted though the score was not (✓). For the remaining five matches Lawro got it wrong.

Lawro made predictions for 371 of the 380 matches in the 2013–2014 season, missing matches which had to be re-arranged because of weather or to accommodate an international match or for some other reason. How did he do? Are Lawro's predictions a good guide to the outcome of a match?

Figure 4.1 shows what happened over the course of the season.

Fig. 4.1 Accuracy of predictions over the season

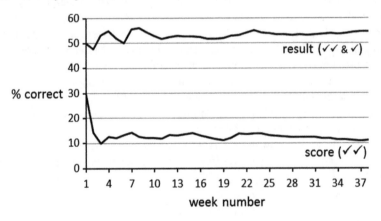

At the start of the season Lawro, like anyone else, had an idea of the relative strengths of the different teams based on an appreciation of their performance during the 2012–2013 season. But some things will be new: new players, new manager, new owner. Three of the twenty teams had been promoted from the lower division of English football so their performance was even harder to predict. As the season progressed results provided an indication of current form and so we might expect predictions to improve. The graph shows the effect of this information. The early season success rate fluctuates quite a lot. As the season wears on the success rate settles down. After week ten Lawro gets about ten percent of scores right but more than half of the results right, fifty-five percent towards the end.

Lawro doesn't become more accurate but he does become more consistent.

But what level of accuracy to measure? That depends on why we want Lawro's help. In a time long ago, before personal computers and pay television, people used to bet on football results through the football pools companies. Each week a punter would select a fixed number of matches and enter them on a form, "the coupon", which would be posted, with a mix of hope and resignation, to the pools company. Points were awarded depending on the results of the chosen matches: three points for a score draw, two points for a no score draw (0–0) and one point otherwise. Points mean prizes, of course. If the total points exceeded some criterion value a prize was given, which could be substantial. With the pools, as with lotteries, most got nothing. In those days all matches were played on Saturdays so that it became a familiar teatime ritual that at five o'clock chatter must cease so that all attention could be given to the radio, later television. Results of the day's matches were checked against the pools entry. "Ah well, better luck next week. We only missed by two points. If only Arsenal had got that other goal."

This ritual is long gone. Punters now have many more online options including placing bets during the match—who will score the next goal?

How useful you find Lawro's predictions depends on why you want them. What is *your* decision? If you are a punter Lawro may help you to decide what the result might be in advance of the match. If you are betting at half-time on the final result Lawro's prediction (made on Thursday, remember) will be of much less use, though you still might want to bear it in mind. If you are betting during the game who will score next, what Lawro said will be of no use at all. This is quite a general point: expert opinion is of use only in the context of a decision. The nature of the decision will determine how useful is the advice, sometimes a lot but sometimes there is no help, you are on your own.

For a non-betting fan the question is simpler: what will happen? Will we win? From this perspective it might be useful to think of one of three match results; an away win (AW), a home win (HW) and a draw (D). How did Lawro do? Table 4.2 shows forecast and result for the 371 matches of the 2013–2014 season.

		Lawro says			
		AW	HW	D	
	AW	43	48	30	121
match result	HW	14	129	30	173
	D	9	37	31	77
		66	214	91	371

Table 4.2 Predictions and results for the whole season

For the whole season he correctly forecast 203 of the 371 results, a fifty-five percent success rate.

Looking at the row and column sums Lawro predicted a lot more home wins than happened and a lot fewer away wins.

We can use the data in Table 4.2 to get likelihood distributions. Just scale each row to sum to one hundred percent (Table 4.3).

alternatives:	evidence: Lawro says			
match result	AW	HW	D	
AW	35	40	25	100%
HW	8	75	17	100%
D	12	48	40	100%

Table 4.3   Likelihoods for using Lawro's predictions

Looking down each column the maximum likelihood corresponds to the actual match result. For any match Lawro is more likely to predict the correct match result than either of the two other possibilities. But how helpful are the predictions? Suppose that Lawro predicts a home win, what should you believe about the result? Use the Bayes Grid in Fig. 4.2.

Fig. 4.2   Bayes Grid analysis for a home win prediction

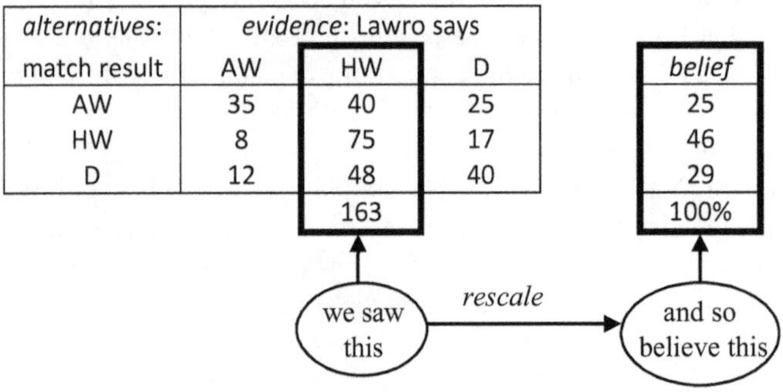

Based only on Lawro's prediction you should believe that a home win is the most likely outcome but the probability of this is just forty-six percent, a little under half. A draw is the next most likely at twenty-nine percent and an away win the least likely at twenty-five percent.

Repeat this analysis for the other possible predictions, an away win or a draw, to see what you should make of them. Table 4.4 shows the justifiable degree of belief you should have in each of the three possible results based only on the evidence of Lawro's prediction.

alternatives:	evidence: Lawro says		
match result	AW	HW	D
AW	64	25	30
HW	14	46	21
D	22	29	49
	100%	100%	100%

Table 4.4 What you should believe will be the result given Lawro's prediction

A predicted away win is the most useful; you should believe that correct with a probability of sixty-four percent.

None of these predictions is perfect. You still have a risky decision to make.

~ • ~

You might have some judgement yourself, independently of what Lawro says, based on your own evaluation of the relative abilities of the teams. How should this be used? You could simply take the probabilities from the table and adjust them according to your judgements: increase this a bit, reduce that a bit and so on. These adjustments may be based on careful consideration or they may be the wishful thinking of the die-hard fan. Does it matter? Not in lively conversation with other fans but you'd need to be careful if you intended to place a big bet. There is a fairly simple way of taking account of your view but making sure that view is well-founded and useful rather than delusional fanspeak is the issue. Wait for Chap. 12.

~ • ~

For some folks there are more important things than football; GDP, for instance. Governments, banks, consultants and others routinely make forecasts of the likely future state of the economy of their own country and sometimes of others where that might be useful. The Organisation for Economic Cooperation and Development (OECD) and the International Monetary Fund (IMF) are two multinational organisations which do just this.

In the case of the IMF one year ahead forecasts are made in October and published in *World Economic Outlook*. These forecasts are numerical, they give an estimate of the future value of GDP. Rather than needing a numerical estimate to be used in a calculation many users of these forecasts want a more general view of whether the economic situation will improve or not, will GDP increase or decrease? The way that a business makes decisions is by discussion and in that context what is most naturally useful is this general view, qualitative rather than quantitative, up or down.

In 2000 Jordi Pons of the University of Barcelona analysed the accuracy of forecasts made by the OECD and the IMF for the G7 group of advanced economies [1]. The seven countries in the G7 at that time were

Canada
France
Germany
Italy
Japan
United Kingdom
United States

For each country he looked at the twenty-two forecasts made between 1973 and 1995. We'll just look at some of the IMF forecasts. For the United States, for instance, Table 4.5 shows the track record.

		IMF forecast		
		increase	decrease	
change	increase	5	3	8
	decrease	3	11	14
		8	14	

Table 4.5 Track record of IMF forecasts for the United States

The IMF mostly got it right. For five of the twenty-two years the IMF correctly forecast an increase in GDP and for another eleven years the IMF correctly forecast a decrease. Sixteen of the twenty-two forecasts, about three quarters, correctly identified the direction of change.

For Japan fourteen forecasts were accurate, a little lower but we have a sample of only twenty-two comparisons (Table 4.6).

		IMF forecast		
		increase	decrease	
change	increase	9	4	13
	decrease	4	5	9
		13	9	

Table 4.6 Track record of IMF forecasts for Japan

Aggregating the twenty-two comparisons for all seven countries gives a sample of 154 of which 111, seventy-three percent, were correct (Table 4.7).

		IMF forecast		
		increase	decrease	
change	increase	52	20	72
	decrease	23	59	82
		75	79	

Table 4.7 Track record of IMF forecasts for the G7

How can an IMF forecast help in deciding what view to take about GDP for the coming year: increase or decrease?

If you were doing this now you would use more recent data, of course, but let's see what you could have done in 1995. Perhaps you are interested in the Japanese economy.

To use the next IMF forecast you must decide what data you will use to establish the track record. Do you want to use just the data for Japan because it is Japan in which you are interested? Or would you rather use data for the whole of the G7 because it provides a bigger sample? Once you've decided that, you will need to think about whether you want to include all twenty-two forecasts. You may think that there has been a significant change in either the Japanese economy or in the IMF methodology or both. But using only the more recent comparisons reduces the sample on which the track record is based. If you aren't sure what to do make more than one analysis and see if you reach different conclusions. Would you make a different decision?

Suppose you decide to consider just the data for Japan but all the data for Japan. As with Lawro's predictions get likelihoods by scaling each row of Table 4.6 to give the likelihood distributions in Table 4.8.

alternatives: GDP change	evidence: IMF forecast		
	increase	decrease	
incre ase	69	31	100%
decrease	44	56	100%

Table 4.8 Likelihood distributions for forecasts of Japan's GDP

The most recent issue of *World Economic Outlook* predicts an increase in Japanese GDP. Now use Bayes Grid in Fig. 4.3.

Fig. 4.3 Bayes Grid analysis for prediction that GDP will increase

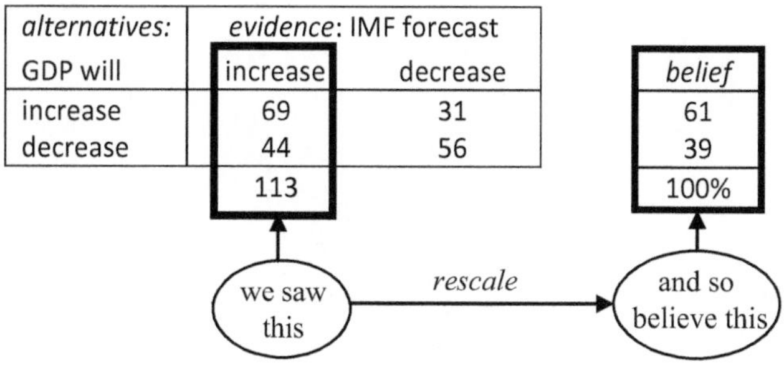

Based on the IMF forecast believe with probability of about sixty percent that Japanese GDP will increase. You may prefer to say that you say there is a 60:40 chance of an increase.

~ • ~

Finding data to establish a track record can be tricky. How useful is past performance in evaluating current advice or evidence?

It is natural to believe that forecasting performance can be improved and is being improved: processes are redesigned, algorithms revised, more data are collected and so on, all in the hope of improvement. Perhaps so. But what is the track record?

There is an apocryphal (I assume) story of a Soviet era Muscovite trying to get a map of the metro. "I'm sorry", says the clerk at the metro station, "we don't have one. We have a map of how it was in the bad old days and a much nicer map of how the metro will look at the completion of the five year plan, but we have no map of how it is today."

If the prediction which you want to use is made only infrequently and not consistently by the same person or method it is hard to establish any sort of track record at all. You may have to accept plausibility rather than accuracy to help you judge. Perhaps you rely on notions of authority or trust in lieu of a track record to help you decide what you should make of the forecast.

Alternatively, you could look at forecasts made by several groups. For example, every month the Treasury (the UK Finance Ministry) publishes *Forecasts for the UK Economy: a comparison of independent forecasts*. In the September 2015 issue there are forecasts from twenty-two City institutions, such as banks, and seventeen non-City organisations including universities and research organisations. You need to note that the *independence* of the forecasts means that the forecasters are independent of government; they are not employed by the Treasury. Whether their forecasts are statistically independent is another matter. We'll get to that in Chap. 11.

~~~ ••• ~~~

References

1. Pons J (2000) The accuracy of IMF and OECD forecasts for G7 countries. J Forecasting 19 (1):53–63

Chapter 5
Game Show

In the cases we have looked at so far data have been central to finding likelihoods. The police evaluating eyewitness evidence used likelihoods based on data found by experiment. Consumers of forecasts should use likelihoods based on data showing track record.

Likelihoods are not always found in this way. Sometimes we have a problem which has the flavour of a puzzle which we can't quite fathom. The structure of Bayes' Rule forces us to think about what we know, but didn't know that we knew. The formalism asks questions which at first might seem awkward but which force us to think clearly. Always a good thing, especially when the solution that looks so obvious is wrong. Our intuition—yours, mine and just about everyone else's—is not always a good guide.

~ • ~

A game show is on television.

It's a very simple game show, even simpler than usual.

A few questions so easy that even though you feel tired after a long day you can answer and feel good about it.

At last, this week's big prize! Will it be two weeks in Florida, a new car, a new house? This week it is a car.

Bob, the host, and Craig, the contestant, stand in front of a wall painted bright pink. In the wall are three doors painted green, yellow and blue. Behind one of the doors, Bob carefully explains (as he has done each week for the years and years that the show has been running), is a car. Behind the other two doors there is nothing. Craig is asked to choose a door.

Craig chooses yellow.

"Are you sure?" asks Bob, "are you really sure?"

Craig does not change his mind.

"OK," says Bob, "now let me help you a little here. Let me make your decision easier. Craig, let me show you what's behind the BLUE door." Bob points to the blue door. It swings open. There is nothing behind the blue door.

"Now Craig, do you want to change your mind?"

Craig looks thoughtful.

© Springer International Publishing AG, part of Springer Nature 2018

A. Jessop, *Let the Evidence Speak*, https://doi.org/10.1007/978-3-319-71392-2_5

Some people in the audience shout "stay with yellow" and others shout "switch to green".

~ • ~

Well, what would you do? And why?

~ • ~

Craig has to make a decision based on evidence.

He has to decide which door he should ask Bob to open. There are three alternatives; yellow and blue and green.

To assist him Bob has helpfully provided some evidence. He has opened one of the doors. He could have opened the yellow door or the blue door or the green door. In this case he opened the blue door.

In the experiment used by the Swedish police the participants, when asked to identify a colour, could choose to answer "blue" or "green". In the game Bob can choose which door to open.

Table 5.1 shows the template for likelihoods. Now we must fill in each row.

alternatives: car is behind	evidence: Bob opens a door			sum
	yellow	blue	green	
yellow				100%
blue				100%
green				100%

Table 5.1 Template for likelihoods

Bob, of course, knows where the car is. If he didn't, he might open a door and show the car. An easy decision for Craig and a guaranteed loss for the show. Bob would soon be looking for a new job.

Suppose the car is behind the yellow door.

This limits Bob's options: he will never open the yellow door and reveal the car. The probability that the yellow door is opened is zero.

As it happens, Craig has chosen the yellow door so Bob can open either of the other doors knowing that the car will not be revealed. There is no reason to believe that Bob favours the blue door or the green door. He is equally likely to open either. Given that the car is behind the yellow door the likelihood distribution describing how likely it is that Bob will open each of the three doors shows zero for the yellow door and equal probabilities for the other two (Table 5.2).

alternatives: car is behind	evidence: Bob opens a door			
	yellow	blue	green	
yellow	0	50	50	100%
blue				
green				

Table 5.2 Likelihood distribution showing what Bob might do if the car is behind the yellow door

Now, suppose the car is behind the blue door.

Bob's choice is even more constrained. He certainly won't open the blue door to show the car.

Neither will he open the yellow door. Craig has chosen this and he needs to be left guessing. Remember that the whole point of this part of the show is to see Craig dithering.

If Bob can open neither the blue door nor the yellow door he *must* open the green door: he has no choice at all. The second likelihood distribution shows this (Table 5.3).

alternatives: car is behind	evidence: Bob opens a door			
	yellow	blue	green	
yellow	0	50	50	100%
blue	0	0	100	100%
green				

Table 5.3 Likelihood distribution showing what Bob might do if the car is behind the blue door

Exactly the same reasoning applies if the car is behind the green door: Bob cannot open the green door to show the car and neither will he open the yellow door which Craig has already chosen. Bob *must* open the blue door (Table 5.4).

alternatives: car is behind	evidence: Bob opens a door			
	yellow	blue	green	
yellow	0	50	50	100%
blue	0	0	100	100%
green	0	100	0	100%

Table 5.4 All three likelihood distributions

We now have a complete description of how Bob's decision, the evidence, is related both to where the car is hidden and to Craig's initial choice. This puts us in a good position to give Craig a helping hand.

Bob opens the blue door.

Use the Bayes Grid in Fig. 5.1 to evaluate this evidence.

Fig. 5.1 What Craig should believe when Bob opens the blue door

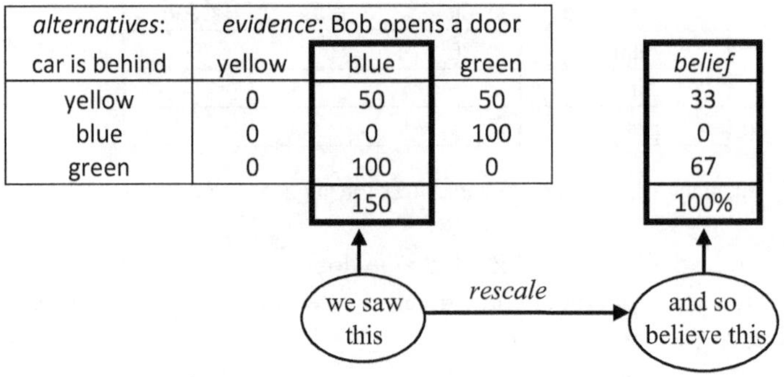

alternatives:	evidence: Bob opens a door				belief
car is behind	yellow	blue	green		
yellow	0	50	50		33
blue	0	0	100		0
green	0	100	0		67
		150			100%

we saw this → *rescale* → and so believe this

Craig should switch. The probability that the car is behind the green door is 67%. There is a probability of only 33% that it is behind the yellow door he originally chose.

This is the general result: the odds are always 2:1 in favour of switching.

~ • ~

But by far the most common response is not to change.

This was my first thought when I saw this problem some years ago. Yours too? Don't worry, we're in the majority.

The most popular argument is that at the start of the show all three doors were closed and the car could be behind any of them: the chance that the car was behind the yellow door, or either of the other two doors, was 1/3. But now one door has been opened. The car is behind one of the two unopened doors and so the chance is reduced to 1/2 for each.

The point here is not that the probability has changed from 1/3 to 1/2 but that in both cases *the probabilities stay the same for all unopened doors*—three at first and then two. There is therefore no reason to prefer one closed door to any other and so there is no advantage to switching from yellow.

But this line of argument fails to recognise the constraint on Bob's decision: he will not, must not, show Craig where the car is. We tend to frame the problem in terms of Craig's decision and fail to recognise that Bob has a decision to make too, and that when Bob decides which door to open he does not have a free choice: in two cases out of three he has no choice at all.

There are other ways to arrive at the correct odds, as your search engine will soon show you, but for me none have the clarity of the Bayesian approach. You are forced to think about likelihoods and so of the process behind them, in this case Bob's point of view and the constraints on his action.

~ • ~

This problem is generally known as the Monty Hall Paradox. It is based on the 1970s American TV show *Let's Make a Deal*. The host was a Canadian called Monty Hall, from Winnipeg (Fig. 5.2). Away from show business he did a great deal of charity work and in 1988, in recognition of this, he was awarded the Order of Canada by the Canadian Government.

Fig. 5.2 Monty Hall[1]

In the show car keys were behind one door and a picture of a goat behind each of the other two.

Deciding the correct answer to the problem—to switch or stick—has generated such useful (and other) discussion over the years that the Monty Hall Paradox is to be found in university departments of mathematics and philosophy, still provoking heat as well as light.

Even if the answer isn't the one you thought of don't be put out; most of us fall into the trap, and that includes professors of maths as well as you and me. For instance, Paul Erdös was a brilliant mathematician and by choice somewhat peripatetic. He used to turn up on the doorstep of the house of a friend or acquaintance with his suitcase and say "Let's write a paper". And so, fuelled by coffee, they did. Erdös wrote lots of papers. But it is said that he went to his grave still wondering what was the correct answer was to the Monty Hall Paradox, and why.

---

[1]Publicity photograph ABC television. Wikimedia Commons. https://commons.wikimedia.org/wiki/File:Monty_hall_abc_tv.JPG. Accessed 18 September 2017.

The correct answer was first given in a popular American magazine, *Sunday Parade*. Marilyn vos Savant (Fig. 5.3) writes a column in which she answers readers' questions. She now also has a website so you can ask too.

Fig. 5.3 Marilyn vos Savant[2]

At one point she was listed in the Guinness Book of Records as having the highest IQ in the world. Just the person to give advice. In 1991 she was asked whether contestants such as Craig should switch. She said that they always should.

And this is where the fun and games started.

Her advice was dismissed by many, including professional mathematicians.

But she was right.

Here's her original answer (taken, as are the responses, from her website[3]):

Yes; you should switch. The first door has a 1/3 chance of winning, but the second door has a 2/3 chance. Here's a good way to visualize what happened. Suppose there are a million doors, and you pick door #1. Then the host, who knows what's behind the doors and will always avoid the one with the prize, opens them all except door #777,777. You'd switch to that door pretty fast, wouldn't you?

[2]Photograph from Shelly Pippin, Quotesgram. Wikimedia Commons. https://commons. wikimedia.org/wiki/File:Marilyn_vos_Savant.jpg. Accessed 18 September 2017.

[3]Marilyn vos Savant. Game Show Problem. http://marilynvossavant.com/game-show-problem/. Accessed 18 September 2017

and here are some of the responses:

> Since you seem to enjoy coming straight to the point, I'll do the same. You blew it! Let me explain. If one door is shown to be a loser, that information changes the probability of either remaining choice, neither of which has any reason to be more likely, to 1/2. As a professional mathematician, I'm very concerned with the general public's lack of mathematical skills. Please help by confessing your error and in the future being more careful.
>
> Robert Sachs, Ph.D.
> George Mason University

> You blew it, and you blew it big! Since you seem to have difficulty grasping the basic principle at work here, I'll explain. After the host reveals a goat, you now have a one-in-two chance of being correct. Whether you change your selection or not, the odds are the same. There is enough mathematical illiteracy in this country, and we don't need the world's highest IQ propagating more. Shame!
>
> Scott Smith, Ph.D.
> University of Florida

> Maybe women look at math problems differently than men.
>
> Don Edwards
> Sunriver, Oregon

So, there you have it. A nice example of the wisdom of self-appointed experts.

~ • ~

The moral to this tale is that we should always consider how data are generated before using them as the basis for a decision.

The apparent paradox in seeing that Craig should switch although it seemed intuitively obvious that this was unnecessary was resolved by a shift in viewpoint from Craig to Bob. Bob's decisions were highly constrained and it was the recognition of these constraints which showed that his apparently whimsical choice contained information which Craig could use. The structure imposed by Bayes Grid made sure that we asked the right questions.

~ • ~

Not all likelihood distributions rely on this sort of reasoning, of course, nor on the tabulation of data in a track record. In a great many cases a mathematical model describes likelihood distributions which have a common form and so are useful for a number of apparently quite different problems. The maths can be a little harder than you want to cope with but the underlying principles are usually easy to grasp, as we'll see in the next chapter.

~~~ ••• ~~~

Chapter 6
Margin of Error

The Stockholm police were able to link Ingrid's eyewitness evidence with the cars of the two suspects because they had likelihoods to help them do it. These likelihoods had been found by simple counting. The experimenters counted how many people said "blue" and how many said "green" when shown a particular colour. There was no theoretical model which could have been used based on the relation between wave length and some visual and cognitive processes which resulted in the probability that a person would name the colour as blue or green. When no theory exists we count.

Which is just what we did to use the track record of Lawro and the IMF.

Being made to specify likelihoods is a useful discipline which sometimes helps us avoid falling into the trap of believing what seems obvious. Sometimes what seems obvious is also correct. Sometimes not, as we found in the game show.

But sometimes we are able to use a little theory to find likelihoods. This is most often possible when what interests us is described by number rather than by categories (blue/green, increase/decrease, switch door/don't). Market researchers want to know what proportion of consumers would buy their new product were it to be introduced. Opinion pollsters want to know the percentage of voters intending to vote for one of the parties in a forthcoming election. People are asked. But not everyone. Neither time nor money are available for that exercise, and so a sample is taken. The results of the survey provide evidence. Bayes' Rule will show what it is then justifiable to believe and how strongly we should have that belief.

~ • ~

In 2015 people living in Scotland voted in a referendum whether they wished to become independent of the United Kingdom. Forty-five percent voted in favour, the Yes vote, and fifty-five percent voted against, the No vote. Independence is the founding principle of the Scottish National Party, SNP, and its predecessors. On becoming the majority party in the Scottish parliament the SNP was committed to a referendum. The result was closer than most commentators expected.

The vote has been lost, this time, but the popularity of the SNP was greatly increased during the campaign and there exists a general feeling that another referendum will be held provided the SNP remains in power in Scotland. A long series of referenda may lie ahead (the "neverendum"). Voter sentiment will be affected by events: the revenue from North Sea oil falls, the fortunes of the SNP and other political parties, Brexit, and many others, no doubt. But feelings for and against the Union run deep and strong. Opinion polls are commissioned by newspapers and others to keep track of popular sentiment.

Between 9 March and 14 March 2017 the polling organisation YouGov asked 1028 people this question

> *If there was a referendum tomorrow on Scotland's future and this was the question, how would you vote? Should Scotland be an independent country?*

and got the answers shown in the first column of Table 6.1.[1]

answer	response: full sample	response: Yes & No only
Yes	37	43
No	48	57
Would not vote	5	-
Don't know	11	-
	100%	100%

Table 6.1 Voting intentions in Scotland

Thirty-seven percent said they would vote Yes

Disregarding the Wouldn't vote and Don't know responses, forty-three percent of those answering either Yes or No said they would vote Yes. This is the headline figure that was seen in the news.

But this is a report based on a sample, not a poll of all voters. Surely, this cannot mean that exactly forty-three percent of all those eligible to vote would vote Yes.

~ • ~

Like the popular media reports of the YouGov survey, let's consider only the answers Yes and No.

The result may be that forty-three percent of the people surveyed said they would vote Yes this time. But we haven't asked everyone. It seems reasonable to believe that if we had then "about" forty-three percent would have said Yes. But what does "about" mean? We might say that forty-three percent intend to answer Yes within a margin of error. The phrase *margin of error* is more helpful than "about" because it implies that some quantitative measure of imprecision is both possible and useful.

Here's how it works.

---

[1]Data given on the YouGov website at https://d25d2506sfb94s.cloudfront.net/cumulus_uploads/document/dj1huiytu2/Times_Scotland_Results_170314_Indy2_Website.pdf. Accessed 18 September 2017.

Keep it simple to start with. First, we have a view that in the population as a whole the proportion who say they would vote Yes is either 40%, 45% or 50%, an unrealistically restricted set of possible alternatives which we'll relax later. Second, we pick one person, Alex, at random and ask him the question. How can we use what he says?

Use the Bayes Grid.

Alex has been picked at random. We know nothing about him and so have no reason to think he is atypical, no reason to think he is more or less likely to vote Yes than anyone else. So if it is true that forty percent of the population say they will vote Yes then the probability that any one of them, Alex in this case, will vote Yes is also forty percent. The probability that Alex will vote No must be sixty percent (our sample excludes the undecided). Table 6.2 shows the three likelihood distributions, one for each of the three alternative proportions of the population who would say they would vote Yes.

alternatives: Yes in population	evidence: Alex says		
	Yes	No	
40%	40	60	100%
45%	45	55	100%
50%	50	50	100%

Table 6.2 Likelihoods for Alex's evidence and three population proportions

Alex says Yes. Use Bayes Grid to find the justified belief we should have about the population's voting intentions (Fig. 6.1).

Fig. 6.1 Bayes Grid shows justified belief about voting intentions

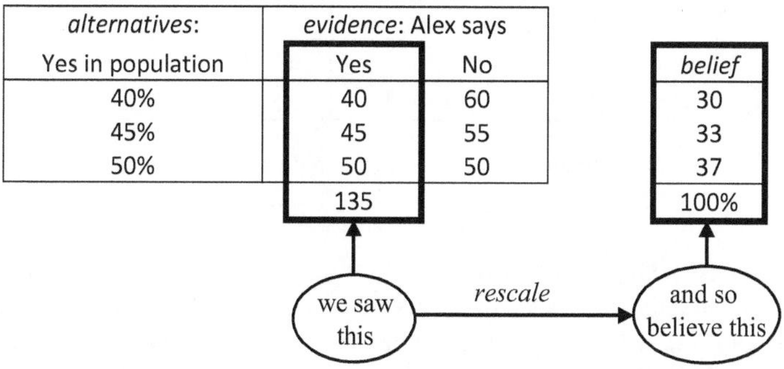

Because Alex says Yes it is natural to believe more in higher values of the population proportion. The Bayes Grid shows just how this happens. We are justified in believing that the probability that forty percent of the population say they will vote Yes is 30%, that forty-five percent say they will vote Yes is 33% and that half say they will vote Yes is 37%.

~ • ~

What happens if we interview two people?

The second person we interview, Nicola, is also chosen at random.

Now we have a sample of two (!) we need a new likelihood distribution.

When we were only considering what Alex might say there were just two possibilities: Alex might say either Yes or No.

Now, with Nicola to consider as well, there are four possible outcomes

Alex says Yes & Nicola says Yes
Alex says No & Nicola says No
Alex says Yes & Nicola says No
Alex says No & Nicola says Yes

What if fifty percent of the population would say they will vote Yes, and so fifty percent would say No?

As before, we have no reason to believe that either Alex or Nicola differ from the population as a whole in what they will say. We also assume that what Alex says is independent of what Nicola says. There is no link between them. Knowing what Alex said provides no information about what Nicola might say, just as knowing what happened when you first toss a coin tells you nothing about what might happen when you toss it a second time.

All four of the outcomes are equally likely. The probability of each is ¼.

Another way of thinking about this is by analogy with tossing a coin twice. The probability of each toss showing heads is ½. To get the probability of two heads just multiply: ½ × ½ = ¼.

The probabilities for each of the four possibilities are shown in Table 6.3.

Alex says	Nicola says	probability
Yes	Yes	0.5×0.5 = 0.25
No	No	0.5×0.5 = 0.25
Yes	No	0.5×0.5 = 0.25
No	Yes	0.5×0.5 = 0.25

Table 6.3 Probability distribution for each of the four outcomes: who says what matters, population proportion = 50%

But we aren't interested in what Nicola might say and what Alex might say as individuals, we only want to know the likelihood of what they both might say taken together: might they both say Yes or might they both say No or might one say Yes and one say No? This last result could happen in two ways, Alex says Yes and Nicola says No or vice versa. Both give the same result, one Yes and one No. We don't care who said what, only how many said what, so we need to add the two probabilities (Table 6.4).

Alex says	Nicola says	probability	number Yes	probability
Yes	Yes	0.5×0.5 = 0.25	2	0.25
No	No	0.5×0.5 = 0.25	0	0.25
Yes	No	0.5×0.5 = 0.25	1	0.5
No	Yes	0.5×0.5 = 0.25		

Table 6.4 Probability distribution for each of three outcomes: who says what irrelevant, population proportion = 50%

The three probabilites in the final column are the likelihood distribution. This shows the probability of what any two people will say if it is true that in the population fifty percent will say Yes.

~ • ~

Was that easy or hard?

In 2013 the Royal Statistical Society and the Institute and Faculty of Actuaries commissioned YouGov to find how many people could correctly answer this question

> *If you spun a coin twice, what is the probability, expressed as a percentage, of getting two heads?*

Fewer than half senior managers and professionals gave the right answer. In a similar survey in 2012 forty-six percent of Members of Parliament got it right [1]. If you had to think a bit you are in good company.

~ • ~

Exactly the same calculation can be used for any other assumed proportion. For example, if the assumed proportion of those in the population saying Yes is forty percent the probability that Alex says Yes and Nicola says No is 0.4 × 0.6 = 0.24.

Find the three probabilities for the likelihood distribution as before (Table 6.5).

Alex says	Nicola says	probability	number Yes	probability
Yes	Yes	0.4×0.4 = 0.16 .	2	0.16
No	No	0.6×0.6 = 0.36	0	0.36
Yes	No	0.4×0.6 = 0.24	1	0.48
No	Yes	0.6×0.4 = 0.24		

Table 6.5 Probability distribution for each of three outcomes: who says what irrelevant, population proportion = 40%

This useful little model is called the Binomial distribution. Use it to get the three likelihood distributions, one for each of the possible population values, as shown in Table 6.6.

alternatives:	evidence: number Yes			
Yes in population	0	1	2	
40%	36	48	16	100%
45%	30	50	20	100%
50%	25	50	25	100%

Table 6.6 Likelihoods for evidence from both Alex and Nicola

We ask Nicola and she says Yes, as did Alex. Of the two interviewed both said Yes. Figure 6.2 shows what we should believe about the population's intention.

Fig. 6.2 Bayes Grid shows justified belief about voting intentions with two Yes votes

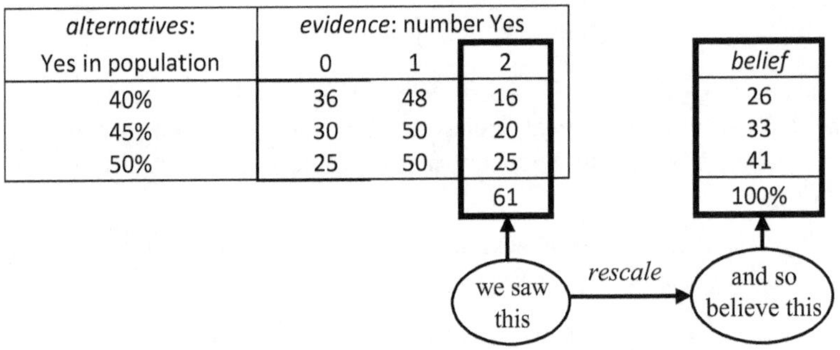

alternatives:	evidence: number Yes			belief
Yes in population	0	1	2	
40%	36	48	16	26
45%	30	50	20	33
50%	25	50	25	41
			61	100%

we saw this → *rescale* → and so believe this

This second Yes should have reinforced our belief that more rather than fewer members of the population will answer Yes. It does, as you can see in Table 6.7.

alternatives:	evidence: Yes from	
Yes in population	Alex	Alex & Nicola
40%	30	26
45%	33	33
50%	37	41
	100%	100%

Table 6.7 Effect of more evidence favouring Yes

If both said No then, naturally, the probability that lower population proportions are the true proportions is increased (Fig. 6.3).

Fig. 6.3 Bayes Grid shows justified belief about voting intentions with two No votes

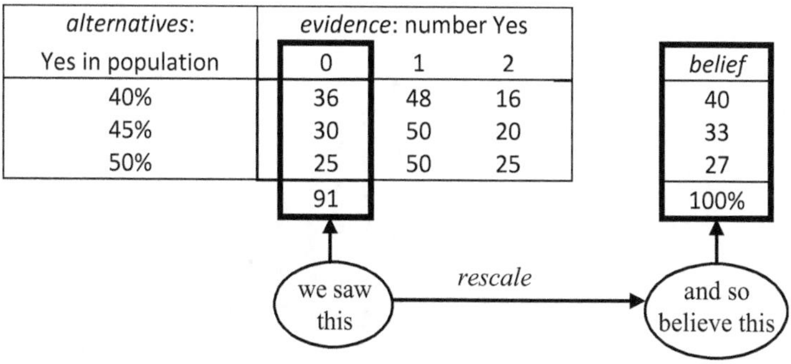

alternatives:	evidence: number Yes			belief
Yes in population	0	1	2	
40%	36	48	16	40
45%	30	50	20	33
50%	25	50	25	27
	91			100%

we saw this → *rescale* → and so believe this

Again, more data (from Nicola) has resulted in more discrimination, this time in favour of the lower values (Table 6.8).

alternatives:	evidence: No from	
Yes in population	Alex	Alex & Nicola
40%	37	40
45%	33	33
50%	30	27
	100%	100%

Table 6.8 Effect of more evidence favouring No

~ • ~

But how ridiculous, you say, this is inadequate for two reasons: the possible population percentages are limited to just three values and a sample of just two people is hopeless.

You're right, of course. This was just a very simple application of the model kept small so that you could easily see the mechanics. Both objections are easily overcome.

Any spreadsheet will have a function which will calculate Binomial probabilities. It will also easily enable you to make a grid which has as many rows and columns as you want. For example, if eight out of twenty people, forty percent, say Yes and we use as possible population percentages 0,5,10... then the estimate, the justified belief, we should have about the population percentage is described by the probability distribution in Fig. 6.4.

Fig. 6.4 What to believe with a sample of 20 (calculated at 5% increments)

This uses exactly the same principles as were used to deal with just Alex and Nicola. The calculation needs more effort but only because more detailed descriptions have been used. The principles are the same: Bayes Grid with more rows and columns but still with Binomial likelihoods. Make the estimate as fine as you wish. Using increments of one percent rather than five percent gives this finer distribution (Fig. 6.5).

Fig. 6.5 What to believe with a sample of 20 (calculated at 1% increments)

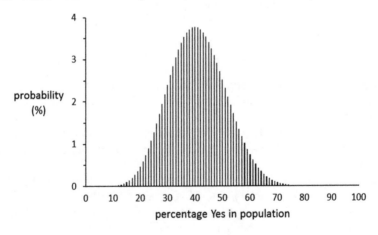

~ • ~

Now we can think again of Bayes' original problem

> *Given* the number of times in which an unknown event has happened
> and failed: *Required* the chance that the probability of its happening

in a single trial lies somewhere between any two degrees of probability
that can be named.

which for us means

Given   that eight people out of twenty said Yes
Required   the chance that the probability that some other member of the
             population will say Yes lies somewhere between any two
             limits.

We do not often want to know the chance that the probability that one person
will say Yes is within a given range. It is much more likely that we want an estimate
of the proportion of the population with that intent. For example, we might want to
find the probability that in the population the proportion saying Yes is within ten
percentage points of that forty percent. The calculation is the same but the interpretation different

Given   that eight people out of twenty said Yes
Required   the probability that in the population the proportion that will
             say Yes lies somewhere between thirty percent and fifty
             percent.

This is just the sum of all the probabilities in that range, the dark bars in Fig. 6.6.

Fig. 6.6 Interval estimate using 1% increments

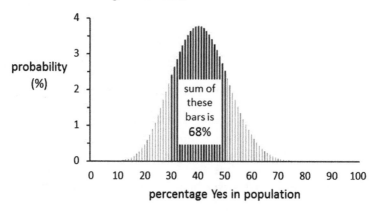

Based on the evidence of our sample of just twenty people we are justified in
believing there is a probability of 68% that between thirty percent and fifty percent
of the population in Scotland will say Yes.

~ • ~

But we can do better than that.

Making the increments smaller and smaller the vertical bars merge.

The result is a smooth curve. The individual bars are indistinguishable. The sum of the bars becomes the area under the curve (Fig. 6.7).

Fig. 6.7 Interval estimate

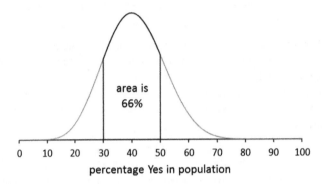

The curve is called a Beta distribution. No new principles have been used, just a bit more maths. Your spreadsheet should have a function to help.

The area under the curve between the limits of thirty percent and fifty percent is 66%. The value of 68% was an approximation, though not too bad.

~ • ~

This model is useful for estimating a percentage from a sample. Market researchers use it to find market share or brand recognition. Engineers use it to estimate the defect rate on a production line from a test sample. You can think of a lot more applications.

What we do not have is a prediction of the referendum.

What we do have is a suitably expressed estimate of the proportion who *say* they would vote Yes. But an opinion poll is not the real thing.

Our estimate is suitably expressed because it gives an indication of the (im)precision of what can be believed from the survey, the margin of error.

~ • ~

The result of such a small survey is usually thought to be too imprecise to be of much use. Increasing the sample size to 500 helps, as Fig. 6.8 shows.

Fig. 6.8 More data give a more precise estimate

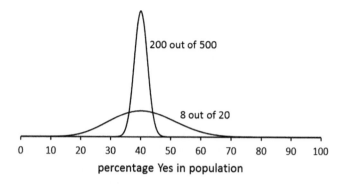

You can now be just about certain that the population percentage is between thirty and fifty.

The probability that the population percentage is within two points of forty percent is 64%. Believe there is a sixty-four percent chance that the population percentage is within two points of forty percent. The margin of error is two percentage points

~ • ~

What was the margin of error of the YouGov poll?

Remember that the sixteen percent who answered either that they would not vote or the they did not know how they would vote were disregarded. The reduced poll result was that forty-three percent of a sample of 864 said they would vote Yes. Figure 6.9 shows the Bayes estimate for the result.

Fig. 6.9 Ninety-five percent interval estimate for YouGov poll

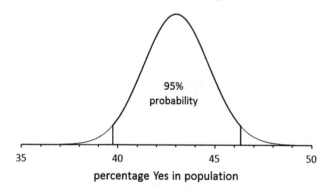

There is a ninety-five percent chance that between forty percent and forty-six percent of all voters would say Yes. This kind of reporting interval is called a *confidence interval*[2]. A ninety-five percent confidence interval is the most popular report to make.

We could also have reported the result as 43 ± 3%, where 3% is a margin of error.

~ • ~

When planning a survey pollsters and others specify a margin as a requirement and then use this to help plan their survey. In the UK it is quite common that samples of about 2000 are used. This gives a probability of ninety-five percent that the true population percentage is within about two points of the survey result, an arbitrary but widely accepted standard. The YouGov survey used a smaller sample and so the margin of error was larger.

As you would expect, the more data you have the smaller the margin of error. Figure 6.10 shows the effect.

Fig. 6.10 More data reduces the margin of error

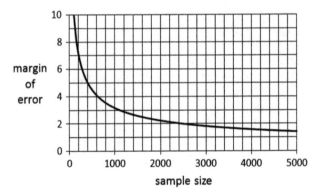

This is the familiar story: to get a better product you must pay more, but see how diminishing returns set in. To halve the margin of error from two percentage points to one point would need a much bigger sample of 8000 people.

~ • ~

<hr>

[2]A confidence interval is symmetric. The tail areas are equal so that the probability that the true value is smaller than the interval is the same as the probability that it is higher than the interval; both 2.5% in this case. You may also read of a *credible interval* or *highest density region*. These intervals are chosen so that the width of the interval is minimised. For symmetric distributions the credible interval is the same as the confidence interval but for skewed distributions it is not. Bayesian statisticians usually favour credible intervals.

In this chapter we have seen how the use of statistical models (the Binomial and Beta distributions in this case) enable an analysis of a standard problem, one that is analytically the same even though the contexts vary

> percentage having a particular opinion
> percentage of defective items
> percentage of satisfied customers
> percentage of patients with an illness

and so on. All have a *parameter* of interest to us, the proportion in the population of voters or customers or whatever. To estimate the value of the parameter we use *evidence* from a sample. This means that the estimate cannot be exact. The Bayesian analysis shows just how inexact, the margin of error.

~~~ ••• ~~~

Reference

1. Royal Statistical Society (2013) RSS research finds that senior execs lack statistical knowledge. Statslife. https://www.statslife.org.uk/news/937-rss-research-finds-that-senior-execs-lack-statistical-knowledge. Accessed 18 September 2017.

Chapter 7
Five Thoughts About Likelihoods

To round off this section on likelihoods and how to use them this chapter shows how thinking in likelihoods can be helpful even when no numerical calculation is used.

The cases we've looked at so far have shown how likelihoods are the basis for deciding what you should believe given some evidence. They determine the strength of belief you should have in alternative causes or explanations of the evidence using Bayes' Rule

belief is proportional to likelihood

If the likelihoods can be described using numbers — probabilities — then so too can the strength of the belief you should have in alternative explanations. For the most part this is straightforward. But even if the analysis is made without numbers the ideas used by Bayes provide a helpful framework for thinking about alternatives and the likelihoods that go with them.

~ • ~

In October 2001 the United States, and much of the world, was trying to decide just what were the implications of the 9/11 attacks on the World Trade Centre in New York. To say that people were anxious would be a colossal understatement. What did it mean? What happens now? What should I do? What should the US do? Reactions ranged from cool appraisal to outright panic. Might there now be a threat from chemical attack?

On 2 October Robert Stevens, who worked for a newspaper in Florida, was admitted to hospital with anthrax [1]. He died three days later. On 9 October, after a second worker in a newspaper mailroom contracted anthrax, the FBI took over the investigation. Since both victims had worked in a mailroom perhaps the anthrax spores came by mail. Letters addressed to Senators Tom Daschle and Patrick Leahy were found to contain anthrax spores. The letters had been posted in New Jersey and processed at the facility in Hamilton Township. They then went to the facility at Brentwood, Washington D.C. The Capitol shut down.

The next day a postal service worker in Hamilton tested positive for anthrax. And the day after that one of his colleagues tested positive as well.

© Springer International Publishing AG, part of Springer Nature 2018
A. Jessop, *Let the Evidence Speak*, https://doi.org/10.1007/978-3-319-71392-2_7

Experts at the Center for Disease Control assured postal officials that envelopes carrying anthrax posed no threat to postal workers. But they were wrong. The buffeting the mail took as it passed through the automatic sorting machines released the micron sized spoors which were then dispersed by the air conditioning system.

On 17 October, a Wednesday, Thomas Morris Jr, who worked at the Brentwood facility in Washington, did not feel well. His doctor diagnosed a virus and sent Morris home. On Sunday his breathing became very laboured and he thought he might have contracted anthrax. He was taken by ambulance to hospital where he died shortly afterwards.

On that same Sunday Joseph Curseen, also a postal worker at Brentwood, complained of flu-like symptoms and was taken to the emergency room of the local hospital. He was sent home. But his symptoms became more acute and so he returned to the hospital. Six hours later he was dead from anthrax inhalation.

Two more victims for the terrorists. Well, no.

In 2004 the FBI believed that the strain of anthrax used in the attacks was RMR-1029. This had been developed in 1997 at USAMRIID (United States Army Medical Research Institute of Infectious Diseases), the US Army's main bioweapons laboratory, by Bruce Ivins. Ivins became the main suspect [2]. He was no terrorist; he had a more personal motivation.[1]

Ivins committed suicide in July 2008.

The FBI closed the case in 2010, concluding that Ivins was guilty.

What has all this to do with likelihoods?

In all, four postal workers at Brentwood had similar symptoms. Two were correctly diagnosed. The doctors who misdiagnosed Morris and Curseen as having a virus admitted later that they did not suspect anthrax until, too late, they had seen the media reports of the other two cases [4].

Evidence, no matter how good, will not help you find something if you're not looking for it. Doctors make diagnoses based on tests and experience. They use this evidence to decide the illness from which their patient might be suffering, if any. Sometimes a test is used to decide whether a patient has a specific illness or not, sometimes to decide which of a number of possible illnesses. Whichever the procedure the doctor must first have identified that illness or those illnesses as a possible cause of the patient's symptoms. The doctors who first saw the postal workers Morris and Curseen were not expecting an outbreak of anthrax in Washington and so did not consider it as a cause for the flu-like symptoms of their patients.

Think back to the Stockholm police. They used eye-witness evidence to help decide which of the two suspects, Stig or Jan, had committed the robbery. But suppose Mats was the robber but was not on the police's list of suspects. Ingrid's evidence won't help find him.

So here is the first point to think about

First thought:
Evidence helps you decide between alternatives. Make sure you have identified all alternatives.

---

[1]The whole sorry tale is told by David Williams [3].

Mostly this will be easy. In a survey of voting intentions alternatives for the percentage intending to vote for a particular party or proposition are all possible values between 0 and 100. But sometimes life is not that simple, so just pause before you move to the next step of finding likelihoods.

~ • ~

If you have to decide how many alternative explanations to use are there any guidelines?

At one level, no. Make sure your set is as comprehensive as possible so that you don't miss anything. You could think of this as the creative phase of the analysis where you can let your imagination be your guide. But the more alternatives, the smaller the belief in any one of them. This may get out of hand.

It's hard to think of any death that has given rise to so many conspiracy theories as the assassination of President John F Kennedy in Dallas on 22 November 1963. Lee Harvey Oswald was arrested for the murder. Two days later Oswald was being led by police through the basement of the building in which he was being held to an armoured car which would take him to the county jail. From the crowd of reporters and assorted onlookers stepped a short man in a suit and a light coloured hat. In his hand was a .38 revolver (Fig. 7.1). He shot and killed Oswald. The killer was Jack Ruby, owner of a Dallas nightclub and a man with underworld connections. This second murder, like the first, was shown live on television.

Fig. 7.1 Jack Ruby shoots Lee Harvey Oswald[2]

[2]Photograph by Ira Jefferson "Jack" Beers Jr., The Dallas Morning News. 24 November 1963. Wikimedia Commons. https://commons.wikimedia.org/wiki/File:Lee_Harvey_Oswald_being_shot_by_Jack_Ruby_as_Oswald_is_being_moved_by_police,_1963.jpg. Accessed 12 September 2017.

How did Ruby fit into this drama? More speculation which, for some, has never been resolved. You can't ask Ruby. He was convicted but appealed the verdict. While awaiting a retrial he died of a pulmonary embolism due to lung cancer. It was January 1967. What to make of it all? Despite the best efforts of the Warren Commission and others the true explanation is still not clear and never will be.

Among the many who have written of these events is the novelist Norman Mailer. It is easy enough to construct a hypothesis to explain Ruby's involvement. Perhaps the CIA had hired the Mafia to kill Fidel Castro and didn't want Oswald to go on trial in case this became public. Perhaps Kennedy got messages about this from Chicago mafia boss Sam Giancana via Judith Exner Campbell, who was friendly with both men. Giancana was alleged to have CIA connections. Is any of this true? As Mailer wrote, "[an] hypothesis, no matter how uncomfortable or bizarre on its first presentation, will thrive or wither by its ability to explain the facts available" [5]. These hypotheses (what Mailer, in another context, called factoids [6]) paint the picture, one of many, about something which inevitably must remain unknowable. The problem is that if there are precious few facts (evidence) these hypotheses might just hang forever in the air until interest wanes.

Second thought:
*The more alternatives you have the smaller the belief in each unless
the evidence sharply discriminates between them.*

But not all alternatives may be as implausible as they first seem.

~ • ~

In his 1955 book *Fact, Fiction and Forecast* the American philosopher Nelson Goodman introduced the world to grue.

We have been used to think of emeralds as green. All the emeralds we have ever seen have been green and so we expect that all emeralds we see in future will also be green. Or they may be grue. If emeralds are grue they will have been green so far but at some point in the future, say next Christmas Day, they will turn blue. Figure 7.2 shows what might happen.

Fig. 7.2   Grue

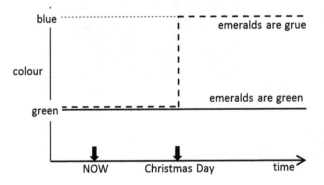

Goodman proposed this as "the new riddle of induction". No past data nor any we collect in the near future can be used as evidence to reject the idea that emeralds are not green, they are grue. As you might expect, there has been much discussion of this grue hypothesis: it seems implausible (to say the least) but where is the rational argument for rejecting it? Is grue a philosophical device to illustrate that some hypotheses are inadmissible, or are there are instances of its existence?

There are. Here are two.

The first example comes from meteorology. Edward Lorenz was a meteorologist who became interested in the unpredictable behaviour of complex non-linear systems, what we now know as chaos theory. He gave a talk at the 139th meeting of the American Association for the Advancement of Science held in Washington, D.C. in December 1972. The title of his talk was *Predictability: Does the Flap of a Butterfly's Wings in Brazil Set a Tornado in Texas?* The Butterfly Effect has joined the Black Swan in the lexicon of science-ish metaphors (The now famous butterfly in the title was actually introduced by the convener of the session, unable to reach Lorenz at the time of the release of the programme [7]).

Lorenz's main point was that the state of the weather system, and so the accuracy of forecasts, can be affected by small, practically imperceptible, changes in some initial conditions used by the forecasting models. You can easily see how this works on your spreadsheet. We are interested in some quantity that varies through time. Its value depends only on the previous value in a fairly simple way: take the previous value, square it, multiply by 2 and subtract 1. How to begin? What is the initial condition? Starting with a value of −0.5 all subsequent values are also −0.5. This stable system is shown in Fig. 7.3.

Fig. 7.3 System behaviour. Initial condition −0.05

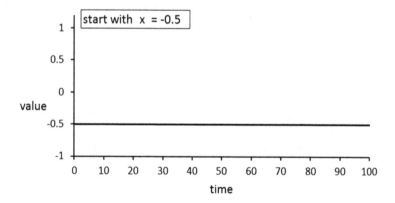

But if the starting value is almost but not exactly −0.5 the system behaves quite differently. After a stable period there is an abrupt switch to instability (Fig. 7.4).

Fig. 7.4 System behaviour. Initial condition almost −0.05

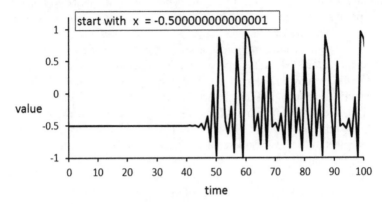

Monitoring values up until, say, the twentieth time period provides no evidence as to which of these two conditions (weather systems) we are experiencing, one that is stable or one that is not.

This simple demonstration is possible because of how computers store numbers. The repeated squaring in the model means that tiny differences in the starting value are likely at some point to have an impact. Now, you may object that all this illustration shows is the potential fragility of computer calculation; if you were doing the calculation by hand it wouldn't happen, you would use −0.5 in both cases. Well, ok, but computer models do behave like that and so might complex systems. Lorenz thought so.

The second example is from economics. The Phillips Curve is named for the British economist AW Phillips. He used data from 1860 to 1957 for his 1958 paper which showed a negative relation between inflation and unemployment. The implication for macroeconomic policy was that governments could adopt policies aimed at reducing unemployment or reducing inflation but not both. But life was not so simple and there was much discussion and research concerning which variables were best to use, whether there is just one curve or many, whether the whole idea made sense and so on. In 1991 FM Akeroyd proposed that the fate of Phillips' original provided an example of grue-like behaviour [8]. Figure 7.5 shows the argument.

Fig. 7.5 Akeroyd's example of grue-like behaviour[3]

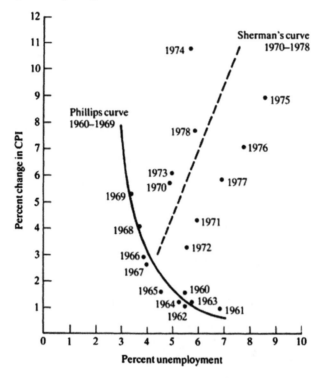

The vertical axis shows change in inflation measured by the Consumer Price Index and the horizontal axis shows unemployment. The solid line is a Phillips curve fitted to 1960s data after which the economic system behaved very differently, as shown by the dashed line. The change of behaviour is sharp. Akeroyd had seen a grue-like switch.

These examples show that grue-like behaviour exists. It goes by many names — discontinuity, paradigm shift, disruption and so on — but it is not, of course, universal. Far from it. Most systems behave over time in fairly predictable patterns. Assuming that models based, as they must be, on the past are reasonable guides for the short to medium term future is useful. But the shadow of grue will not go away. There may be a sharp switch. We have no way of knowing whether this will happen,

---

[3]Graph courtesy of *The British Journal for the Philosophy of Science*: Oxford University Press [8].

nor to what extent, nor what we might do except put in place some careful monitoring and contingency planning.

But don't panic. This does not mean that forecasting and planning are doomed to failure, just that a careful modesty is appropriate. For the most part forecasts are made and, as we have seen, a track record gives a basis for assessing how they might be interpreted and used.

How does all this effect Bayes? The basic point, that it is impossible to construct a likelihood distribution for the future, is clear and unanswerable. The best we can do, and it is very often good enough, is to establish a track record and assume that the future will be like the past.

Third thought:
It is not possible to give likelihoods for the future. It usually makes
sense to assume that the past is a decent guide to the future but be
aware of this extra assumption you are making.

~ • ~

Grue prompts another thought about likelihoods. Here is the application of Bayes Rule. First, the likelihoods (Table 7.1) and then, once we have looked at lots of emeralds and seen that all are green, use Bayes Grid for the analysis (Fig. 7.6).

alternatives: emeralds are	evidence: we see emeralds		
	green	blue	
green	100	0	100%
grue	100	0	100%

Table 7.1 Likelihoods for deciding about grue

Fig. 7.6 Belief for green and grue

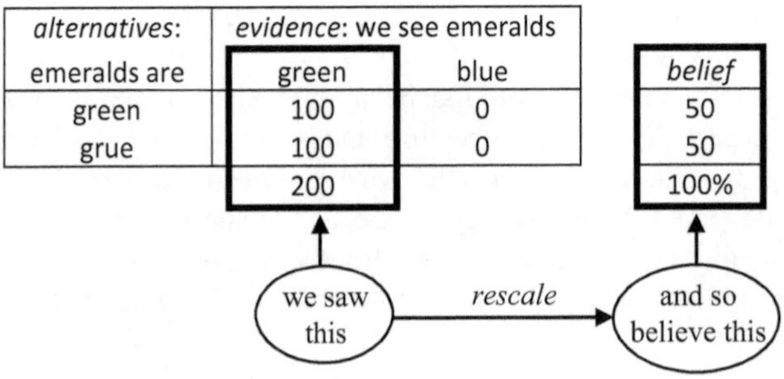

alternatives: emeralds are	evidence: we see emeralds		belief
	green	blue	
green	100	0	50
grue	100	0	50
	200		100%

we saw this   *rescale*   and so believe this

Given the evidence we are unable to differentiate between the alternatives green and grue. We have to believe that both are equally likely because the likelihoods are the same. It is not the values of likelihoods but that they are the same which makes it impossible to use evidence to discriminate between alternatives.

You may, of course, be more sceptical about the plausibility or admissibility of some alternative explanations. You could choose to decide either that alternatives are plausible enough to be included in your analysis or they are so implausible that you rule them out (grue? anthrax?). Perhaps, though, you would like some intermediate option that enables you to give a weight to each alternative according to its plausibility. Some people think this sharp practice. This is an important point that we'll look at later.

More familiar than grue is the "well, she would say that, wouldn't she" problem. Why would a suspect confess to a crime? Some do because of remorse or a recognition of their inevitable conviction but otherwise what is the value of a denial? Almost certainly, nothing.[4] Table 7.2 shows the likelihoods.

alternatives:	evidence: suspect's statement		
suspect is	deny	confess	
innocent	100	0	100%
guilty	100	0	100%

Table 7.2 Likelihoods for confession

Extreme differences in the diagnostic power of evidence are not uncommon. Just how useful data are as evidence depends on how well they can help you discriminate between alternatives.

Suppose that you wanted to know whether a message had been written in English or French but that a system error meant that all you had was one letter retrieved from the damaged email. How useful might that be? It could be decisive. If the letter was accented — ô or é or à, for instance — you would know for certain that the message was written in French. (French words such as cliché are also used in English, but very infrequently.) You are not so fortunate but may still be able to

---

[4]This analysis wouldn't hold everywhere. In the United States the overwhelming majority of cases never get to court. Most accused wish to avoid very harsh sentences and so plead guilty to a lesser charge, of which they might be innocent. In *The Atlantic* of September 2017 Emily Yoffe reports that 97% of federal cases are settled by these plea bargains and so never make it to the courtroom [9].

make use of the different frequency of occurrence of letters in the two languages. For example, Table 7.3 gives the frequencies of occurrence[5] of the letters n, w, x and f.

alternatives:	evidence: character retrieved					
language	n	w	x	f	other	
English	7.17	1.83	0.19	2.18	88.63	100%
French	7.32	0.08	0.43	1.12	91.05	100%

Table 7.3 Likelihoods for letters in English and French

These are two likelihood distributions so Bayes Grid shows what you are justified in believing. If an n is retrieved almost no discrimination is possible (Fig. 7.7).

Fig. 7.7 Belief if the letter n is seen

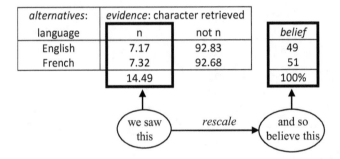

There is only the slightest of evidence in favour of English because the likelihoods, 7.17 percent and 7.32 percent, are almost the same. The other three letters occur much more frequently in one language than the other and so provide more

---

[5]The precise frequency of occurrence of letters in a language is not known and never can be, there are just too many variables such as when the language is used, in what medium it is expressed, by whom and for what purpose. There are many lists showing frequencies found by sampling texts written in that language. The data used here are from Practical Cryptography. Letter frequencies for various languages. http://practicalcryptography.com/cryptanalysis/letter-frequencies-various-languages/. Accessed 4 July 2017.

discrimination, more useful evidence. Table 7.4 shows what happens if you use Bayes Grid for each of the four letters.

alternatives:	evidence: character retrieved			
language	n	w	x	f
English	49	96	31	66
French	51	4	69	34
	100%	100%	100%	100%

Table 7.4 Beliefs depending on which of four letters is seen

The letter w is almost perfectly discriminating. If you retrieve w you are justified in believing very strongly that the message was written in English.

Retrieving either x or f provides about the same not so strong evidence, x in favour of French and f in favour of English. The odds are about 2:1.

Of course, you would want more evidence than just one letter. This example is to illustrate how Bayes thinking can help you decide how to interpret the evidence in just the same way as the sample of just two people illustrated the idea behind margin of error in Chap. 6. To decide between texts you might look at the occurrence of word pairs or triples (digrams or trigrams) or perhaps letter frequencies in a sample of words. We'll look at this again in Chapter 13.

Fourth thought:
Likelihoods help you discriminate between alternatives to the extent
they are different. Not all data provide useful evidence.

The likelihood ratio determines how discriminating is the evidence you have. Should you also pay attention to the values of the likelihoods? Does it matter if the evidence, however discriminating, shows you have seen the commonplace or the unusual or the very rare? It depends on what you want to do with the information.

~ • ~

Interpreting differences in likelihoods when the probabilities are very low is always likely to be tricky. This is especially true where changes in physical risks are concerned. Margaret McCartney is a family doctor in Glasgow. In November 2006 she wrote a piece in the *Financial Times* headed "Lies, damn lies and medical statistics". Her point was that some statistics are unhelpful and some are useful. More particularly, the way some statistics are presented and the way we perceive them makes interpretation not straightforward. Those she did not like included that taking one type of oral contraceptive pill increased the relative risk of a blood clot three times. But that only meant that the risk went from 5 per 100,000 to 15 per 100,000.

Presumably, data had been collected and tabulated according to the presence or absence of a blood clot and also whether or not a pill had been taken. The results can be shown as likelihoods (Table 7.5) from which we get the Bayes Grid analysis (Fig. 7.8).

alternatives: pill?	*evidence:* blood clot? yes	no	
yes	0.015	99.985	100%
no	0.005	99.995	100%

Table 7.5 Likelihoods of blood clot

Fig. 7.8 Blood clot three times more likely if pill taken

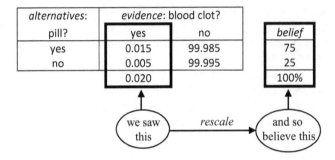

Although the relative risk had trebled the absolute risk was still very small. Which story to tell?

On the other hand, just because likelihoods are low does not mean they can be, or should be, ignored. Think of how much money and effort is used to reduce the already low risk of an aircraft collision. A judgement must be made which takes account both of the likelihood and the severity of the consequences. Different people will reach different conclusions.

Fifth thought:
Differences in likelihoods help you decide between alternatives but
don't lose sight of just how likely the evidence is.

Juries consider even smaller likelihoods when dealing with DNA evidence. This is not always straightforward, as we'll see in Chapter 16.

~ • ~

In this first section of the book we have seen the vital role of likelihoods in thinking rationally about how to use evidence. But evidence collected in one situation may be used in another. If this new context is different how can this difference be taken into account? We now turn to that problem and see how a simple extension of Bayes' Rule will help.

~~~ ••• ~~~

References

1. Pope N (2001) The anthrax mail attack; parts 1 and 2. Pushing the Envelope: Smithsonian's National Postal Museum blog. http://postalmuseumblog.si.edu/2011/10/anthrax-mail-1.html, http://postalmuseumblog.si.edu/2011/10/anthrax-attack-2.html. Accessed 4 July 2017
2. History Commons (2001) Anthrax attacks. http://www.historycommons.org/timeline.jsp?time line=anthraxattacks&anthraxattacks_suspects=anthraxattacks_bruce_ivins. Accessed 4 July 2017
3. Williams D (2011) The mirage man. Bantam Books, New York
4. MacKenzie D (2001) Whodunnit? A loner with a grudge is behind the anthrax attacks, says FBI. New Sci 172(17):10
5. Mailer N (1995) The amateur hit man. NY Rev Books 42(8):52–59
6. Mailer N (1973) Marilyn: a biography. Grosset and Dunlap, New York
7. Rouvas-Nicolis C, Nicolis G (2009) Butterfly effect. Scholarpedia 4(5):1720. http://www.scholarpedia.org/article/Butterfly_effect. Accessed 4 July 2017
8. Akeroyd FM (1991) A practical example of grue. Brit J Philos Sci 42(4):535–539
9. Yoffe E (2017) Innocence is irrelevant. The Atlantic 330(2):66–74

Part II
Base Rate

Chapter 8
Diagnosis

The chapters in the previous section showed how Bayes' Rule uses likelihoods to evaluate evidence and arrive at a justified belief in the alternative explanations for what we see. There may be other contextual factors, not part of this evidence, which we would like to include. Medical diagnosis is a good example. A doctor may want to use both the results of some useful test together with what is known about a particular patient or group of patients; some may be in a high risk group and some not. In this chapter we will see how this might be done.

Data show how well a test result indicates the presence or absence of a condition which, if present, may be improved by treatment. The risks of treatment, surgery most obviously, might well be high but the risk of doing nothing may be higher.

~ • ~

About one in a thousand people in the UK suffer from some form of venous thrombosis or deep vein thrombosis (DVT). A blood clot forms and obstructs the flow of blood. This most often occurs in one of the deep veins of the leg. Doctors and others working in emergency rooms need a quick method of assessing whether a patient has DVT. Given pressures on budgets it would be helpful if such tests were not too expensive.

In 2001 a team at the Prince Charles Hospital in Merthyr Tydfil, in Wales, evaluated three such diagnostic tests [1].

Venography is a procedure in which a dye is injected into the veins and an x-ray image, a venogram, is made so that blood clots may be seen and a sure diagnosis made. The team called this the "gold standard" procedure. But it requires the infrastructure of a laboratory, lab technicians and the rest. It takes time to produce a result. At the time of the study each test cost £30.

D-dimer refers to the small fragments of the protein fibrin in the blood. Concentration is measured in milligrams per litre of blood. Low values indicate a blood clotting system operating as it should. Higher values indicate that all may not be well. By using a reference level the test result is returned as being either negative, low enough that no thrombosis is indicated, or positive. SimpliRED is one of several commercially available testing products used for D-dimer analyses. The team in Merthyr Tydfil were interested in this because it was quick (10 minutes),

© Springer International Publishing AG, part of Springer Nature 2018
A. Jessop, *Let the Evidence Speak*, https://doi.org/10.1007/978-3-319-71392-2_8

cheap (£7), and could be used at the patient's bedside. Just the thing for an emergency department provided it was a good enough indicator of the presence or absence of DVT.

To see whether it was good enough 187 patients with clinical symptoms indicating the possibility of DVT had blood taken. Samples were analysed both by the SimpliRED test and a venogram (the "gold standard"). The results were presented as shown in Table 8.1.

		Venogram result		
		positive	negative	
SimpliRED	positive	38	23	61
test result	negative	13	113	126
		51	136	187

Table 8.1 Comparison of SimpliRED and venogram results

This looks promising: 151 of the 187 results are the same for both tests. SimpliRED may well be useful.

In the medical literature tables show the different test results as rows. In the Bayes Grid, however, test results, the evidence, are the columns. Make that switch here.

Because of the accuracy of the venogram its result can be taken as meaning that DVT really is present or is not and so the data can be taken as showing the relation between the SimpliRED result and the patient's condition (Table 8.2).

		SimpliRED test result		
		positive	negative	
DVT ?	yes	38	13	51
	no	23	113	136
		61	126	187

Table 8.2 Comparative data reformatted

Doctors want to know how accurate the test is and for this use two ideas: *specificity* is the ability of the test correctly to identify healthy patients and *sensitivity* is the ability of the test correctly to identify patients who are ill.

Of the 136 patients with no DVT the test correctly identified 113 and so

$$specificity = 113/136 = 83.1\%.$$

51 patients had DVT of whom the test correctly identified 38 and so

$$sensitivity = 38/51 = 74.5\%.$$

Specificity and sensitivity show how likely is a test result given that we know whether the patient has DVT or not: they are Bayes likelihoods, as shown in Table 8.3.

But the whole point of the test is that we do not know whether the patient has DVT and so the doctor would like to know how well the test performs. To do this it is customary to look at the data and see that

of 61 patients testing positive 38 (62.3%) had DVT,

and   of 126 patients testing negative 113 (89.7%) did not have DVT

which is just the same as rescaling the table of data (Table 8.2) so that the columns sum to 100% (Table 8.4).

		SimpliRED test result		
		positive	negative	
DVT ?	yes	74.5	25.5	100%
	no	16.9	83.1	100%

Table 8.3 Likelihoods of test results

		SimpliRED test result	
		positive	negative
DVT ?	yes	62.3	10.3
	no	37.7	89.7
		100%	100%

Table 8.4 Diagnostic probabilities

These two tables give different views of the data. The first, showing specificity and sensitivity, describes the characteristics of the test. The second shows the predictive power of the test as used at the Prince Charles Hospital. These two views use the same data. To see the link between them use the Bayes Grid.

~ • ~

The likelihoods are just the specificity and sensitivity so if the SimpliRED result was positive the doctor should believe that the probability the patient has DVT is 82% (Fig. 8.1). But this is different from the 62.3% found above (Table 8.4).

Fig. 8.1 Bayes Grid analysis shows what to believe given a positive test result

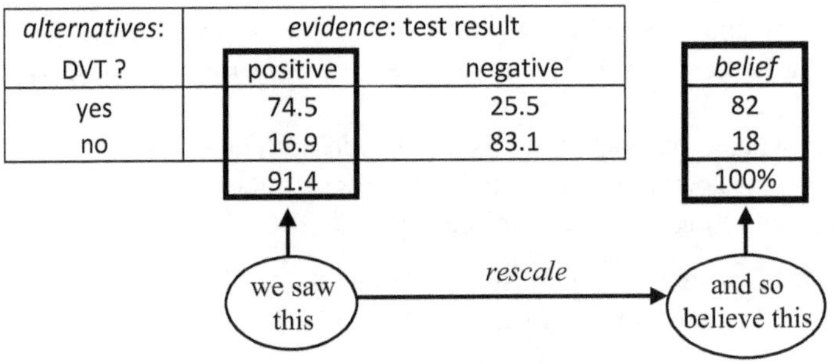

alternatives:	evidence: test result			belief
DVT ?	positive	negative		belief
yes	74.5	25.5		82
no	16.9	83.1		18
	91.4			100%

we saw this → *rescale* → and so believe this

If the test result was negative the probability that the patient does not have DVT is 77% (Fig. 8.2), not the 89.7% of the earlier calculation.

Fig. 8.2 Bayes Grid analysis shows what to believe given a negative test result

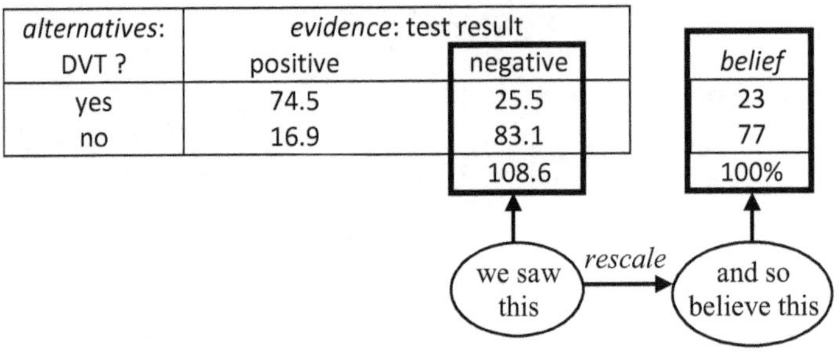

alternatives:	evidence: test result			belief
DVT ?	positive	negative		belief
yes	74.5	25.5		23
no	16.9	83.1		77
		108.6		100%

we saw this → *rescale* → and so believe this

Why these differences? Why does this Bayes analysis not give the same result as that reported by the doctors?

Look at the likelihood distributions. Both rows sum to 100%. In the calculation both are given equal weight. No differentiation is made between the number of patients with DVT and those who don't have DVT. But this is not what the data tell us.

Of the 187 patients in the experiment 51, only 27.3%, actually had DVT. This is called the *prevalence* of the condition. Before using SimpliRED, or any other test, the doctors at Merthyr Tydfil can reasonably believe that 27.3% of their patients, about one in four, has DVT.

This belief is formed before and independently of the diagnostic test and has several names, *prior probabilities* is used a lot. We shall call them *base rates*, which is the name popular with psychologists. These base rates describe what we know of the patient given a particular context: in this case that they are patients in the emergency room at the Prince Charles Hospital and showing symptoms of DVT. They are not necessarily typical of the wider population.

It is natural to want to use both the contextual information and the test result when making a diagnosis and deciding treatment. Do this by weighting the likelihoods by the base rates and then rescaling as before. Table 8.5 shows the calculation.

alternatives: DVT ?	base rate	evidence: positive	base rate × likelihood	belief
yes	27.3	74.5	27.3×74.5 = 2033.85	62.3
no	72.7	16.9	72.7×16.9 = 1228.63	37.7
	100%		3262.48	100%

Table 8.5 Using Bayes' Rule with base rates

The result is just the same as found by the doctors in their original analysis. Bringing in the base rates makes the difference. As Fig. 8.3 shows, Bayes Grid is easily modified to incorporate this reasoning.

Fig. 8.3 Bayes Grid extended to include base rates with positive test result

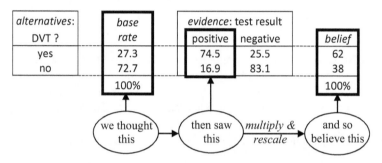

Setting out the Bayes model like this forces you to answer a series of useful questions. Reading from left to right,

what are the alternatives?
what do you know already?
what evidence might you see?
what did you see?
what should you now believe?

This emphasizes that Bayes' Rule is a framework for *learning*, for revising an *initial belief* in the light of evidence:

belief is proportional to base rate × likelihood

This restatement of Bayes' Rule is the one most used. The inclusion of base rates greatly extends the scope and usefulness of Bayesian analysis, as we shall see.

In the case of a negative test result we are justified in believing that there is a 90% probability that all is well, the patient does not have DVT (Fig. 8.4).

Fig. 8.4 Bayes Grid for negative test result

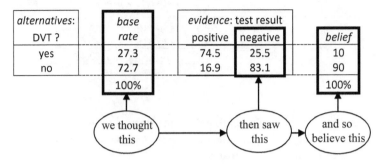

There is still a 10% chance that the patient does have DVT, of course.

~ • ~

The first Bayes calculation used only likelihoods. No use was made of base rates. This is why the result differed from the doctors' result. We can now see that this is just a special case. Without any contextual information the likelihoods are weighted equally which is equivalent to using equal base rates, 50:50 in this case (Fig. 8.5).

Fig. 8.5 Using equal valued base rates when there is no contextual information

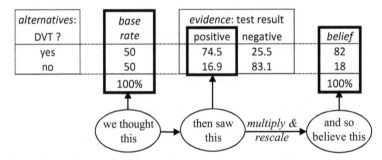

Using equal base rates describes what you ought initially to believe in the absence of any background information. This is an important (if apparently obvious) idea to which we shall return

~ • ~

Setting out the problem in this way makes clear the different contributions of base rate and likelihood in deciding what ought to be believed about a patient's medical condition given the result of a SimpliRED test.

The likelihoods describe the properties of the test, its sensitivity and specificity. The response to some tests depends on the ethnicity or other characteristics of the patient, but for patients with the same characteristics as the Welsh these likelihoods hold whether the test is given for a patient in Merthyr Tydfil or New York or Canberra.

The base rates describe just who is the patient; in a hospital in Wales, at a general practitioner's office in Wisconsin or whatever. The doctors will want to ask not simply

> given this test result how likely is my patient to have DVT?

but   given this test result *and given any other relevant information I have* how likely is my patient to have DVT?

The doctors in Merthyr Tydfil knew that 27.3% of their patients have DVT. This is the base rate. Other factors may be relevant (Fig. 8.6).

Fig. 8.6 Risk factors for DVT[1]

Who is at risk?

Each year, 1 in every 1,000 people in the UK is affected by DVT.

Anyone can develop DVT, but it becomes more common with age. As well as age, risk factors include:

- previous venous thromboembolism
- a family history of blood clots
- medical conditions such as cancer and heart failure
- inactivity – for example, after an operation being overweight or obese

If the patient is known to have a family history of blood clots and also does a lot of long distance flying, and is therefore inactive for long periods, that can in principle be incorporated into the analysis provided data are to hand. Base Rates help to do this.

~ • ~

---

[1](Based on information provided at NHS Choices. http://www.nhs.uk/Conditions/Deep-vein-thrombosis/Pages/Introduction.aspx. Accessed 18 September 2017).

Using Bayes Grid makes very clear the process and the reasoning behind that process. An initially blank Grid asks questions and prompts you to enter all the information needed for the application of Bayes' Rule. The rescaling shows what you should believe once one particular result is seen.

It is sometimes useful to use a different layout which shows the analysis in three stages and for all test results. Figure 8.7 shows how.

Fig. 8.7 Bayes' Rule in three tables

<table>
<tr><td></td><td></td><td></td><td colspan="2">SimpliRED test</td><td></td></tr>
<tr><td>1. What are the likelihoods?</td><td></td><td></td><td>positive</td><td>negative</td><td></td></tr>
<tr><td>sensitivity = 74.5%</td><td rowspan="2">DVT ?</td><td>yes</td><td>74.5</td><td>25.5</td><td>100%</td></tr>
<tr><td>specificity = 83.1%</td><td>no</td><td>16.9</td><td>83.1</td><td>100%</td></tr>
</table>

<table>
<tr><td></td><td></td><td></td><td colspan="2">SimpliRED test</td><td></td></tr>
<tr><td>2. What are the base rates?</td><td></td><td></td><td>positive</td><td>negative</td><td></td></tr>
<tr><td>prevalence = 27.3%</td><td rowspan="2">DVT?</td><td>yes</td><td>20.3</td><td>7.0</td><td>27.3</td></tr>
<tr><td>Rescale rows.</td><td>no</td><td>12.3</td><td>60.4</td><td>72.7</td></tr>
<tr><td></td><td></td><td></td><td>32.6</td><td>67.4</td><td>100%</td></tr>
</table>

<table>
<tr><td></td><td></td><td></td><td colspan="2">SimpliRED test</td></tr>
<tr><td>3. Believe this depending on</td><td></td><td></td><td>positive</td><td>negative</td></tr>
<tr><td>the test result.</td><td rowspan="2">DVT?</td><td>yes</td><td>62.3</td><td>10.3</td></tr>
<tr><td>Rescale columns.</td><td>no</td><td>37.7</td><td>89.7</td></tr>
<tr><td></td><td></td><td></td><td>100%</td><td>100%</td></tr>
</table>

For medical diagnoses the column sums at step 2 may themselves be of interest. Even though sensitivity and specificity are quite high about one third of patients, 32.6%, will receive a positive test result. If you have ignored the prevalence, the base rate, this will seem counterintuitive.

This test is given in an emergency room. Patients with a positive SimpliRED test result will be referred for further investigation. A venogram will confirm the diagnosis of DVT for 62.3% of those referred. The others will be relieved to be told that they don't have DVT after all.

~ • ~

Having a clear structure helps make sense of data. Filling out the Bayes Grid forces you to answer the necessary questions which you might otherwise overlook. In thinking about tests such as the SimpliRED test it is easy to focus on sensitivity and specificity and overlook prevalence, the base rate. We are all prone to this mistake as the following simple experiment shows.

Some children are born with Down's syndrome. There are tests which can be given to pregnant women to see if their baby may suffer from this condition. A team at the University of Liverpool wanted to see how well test results were interpreted by those involved; the pregnant women, their companions, midwives and obstetricians [2]. Eighty-five people were shown this scenario

> The serum test screens pregnant women for babies with Down's syndrome. The test is a very good one but not perfect. Roughly 1% of babies have Down's syndrome. If the baby has Down's syndrome, there is a 90 per cent chance that the result will be positive. If the baby is unaffected, there is still a 1% chance that the result will be positive. A pregnant woman has been tested and the result is positive. What is the chance that her baby actually has Down's syndrome?
>   ____ %

Read this again and write down your answer.

~ • ~

Table 8.6 shows how well the eighty-five people performed.

	correct	overestimate	underestimate	
pregnant women	1	15	6	22
companions	3	10	7	20
midwives	0	10	12	22
obstetricians	1	16	4	21
	5	51	29	85

Table 8.6   Diagnostic accuracy for Down's syndrome

Only five of the eighty-five gave the correct answer. The health professionals did no better that the pregnant women and their companions.
How about you?
Figure 8.8 shows the Bayes Grid.

Fig. 8.8   Bayes Grid analysis for Down's syndrome data

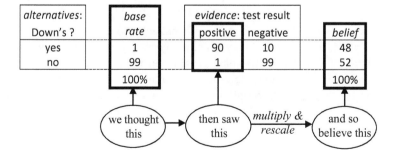

With a positive test result there is a 48% chance that the baby has Down's syndrome.

Using the Bayes Grid forces us to think about base rates and how to use them.

~ • ~

Interestingly, a different group of eighty-one people were shown this alternative scenario:

> The serum test screens pregnant women for babies with Down's syndrome. The test is a
> very good one but not perfect. Roughly 100 babies out of 10 000 have Down's syndrome.
> Of these 100 babies with Down's syndrome, 90 will have a positive test result. Of the
> remaining 9900 unaffected babies 99 will still have a positive test result. How many
> pregnant women who have a positive result to the test actually have a baby with Down's
> syndrome?
> ____ out of ____

The information is the same but now given as numbers of women and babies rather than as percentages. Table 8.7 shows the result.

	correct	overestimate	underestimate	
pregnant women	3	8	10	21
companions	3	8	9	20
midwives	0	7	13	20
obstetricians	13	3	4	20
	19	26	36	81

Table 8.7 Diagnostic accuracy for Down's syndrome using reformulated data

This is clearly an improvement. Why?

The reframing of the scenario makes it an easier problem. All that is required is to pick out the two numbers 90 and 99 and so get $90/(90+99) = 48\%$.

But still only about a quarter of the sample got the right answer.

~ • ~

Diagnosis means deciding the condition of a patient who has gone to a doctor with some symptoms and wants to know what, if anything, is wrong with them.

Screening is a precautionary mass program in which anybody may be tested even though they exhibit no symptoms. Well, often not just anybody exactly. The offer may be made only to people in high risk groups such as older people, for instance.

What would happen if SimpliRED was used in a screening program for all adults? The precision of the test as measured by sensitivity and specificity is unaltered but the base rates change. In the UK about one person in every thousand gets DVT each year. The Bayes' Rule analysis with these base rates is shown in Fig. 8.9.

Fig. 8.9 Using SimpliRED for screening

1. What are the **likelihoods**?

sensitivity = 74.5%
specificity = 83.1%

		SimpliRED test		
		positive	negative	
DVT ?	yes	74.5	25.5	100%
	no	16.9	83.1	100%

2. What are the **base rates**?

prevalence = 0.1%

Rescale rows.

		SimpliRED test		
		positive	negative	
DVT?	yes	0.1	0.0	0.1
	no	16.9	83.0	99.9
		17.0	83.0	100%

3. Rescale columns.

Believe this depending on the test result.

		SimpliRED test	
		positive	negative
DVT?	yes	0.4	0.0
	no	99.6	100.0
		100%	100%

Seventeen percent of those tested would receive a positive result and so would be referred for a further test, a venogram. The overwhelming majority of these, 99.6%, would be cleared. This is a much higher percentage than with the patients in the Prince Charles Hospital because the prevalence of DVT in the population is much less than in the hospital's patients.

Should this screening program be adopted? No. Too many people would be referred for tests which, for almost all of them, would be negative. They would be worried while they were waiting for this second test. And then there's the cost. The prevalence of this condition is too low to justify the expenditure.

In contrast to this, in many countries it is common to screen women for breast cancer by using an x-ray called a mammogram. If the mammogram indicates the possible presence of a tumour the woman is referred for further investigation, a second mammogram or a biopsy for instance. Practice varies in different countries but it is usual to offer breast screening to women older than about forty. Methods and procedures for interpreting results also vary. A recent study [3] in the US found a sensitivity of 83% and a specificity of 92%. Prevalence in the US is about 12%. Figure 8.10 shows what to expect from the screening.

Fig. 8.10 Screening for breast cancer

		mammogram positive	mammogram negative	
1. What are the **likelihoods**?				
cancer?	yes	83	17	100%
sensitivity = 83%	no	8	92	100%
specificity = 92%				

		mammogram positive	mammogram negative	
2. What are the **base rates**?				
prevalence = 12%	yes	10.0	2.0	12
cancer?	no	7.0	81.0	88
Rescale rows.		17.0	83.0	100%

		mammogram positive	mammogram negative
3. Rescale columns.			
Believe this depending	yes	58.6	2.5
on the test result. cancer?	no	41.4	97.5
		100%	100%

About one in six women (17%) will be referred. Of these about 40% will be found not to have breast cancer. You might think this surprisingly high if you have in mind only the sensitivity and specificity values of about 80% and 90% and so expect only about 10% of incorrect diagnoses. But this expectation is false: we have ignored the base rate.

Of women referred about forty percent will be found not to have breast cancer. In the time between screening and result they will have been concerned. This is unavoidable given the accuracy of mammograms and, particularly, the prevalence of the disease. Most women think this a price worth paying.

~ • ~

Base rates are important because they allow you to make use of important contextual information. Using either the Bayes Grid or the three table format forces you to provide *all* the necessary information for Bayes' Rule

belief is proportional to base rate × likelihood

~~~ ••• ~~~

References

1. Neale D, Tovey C, Vali A, Davies S, Myers K, Obiako M, Ramkumar V, Hafiz A (2004) Evaluation of the Simplify D-dimer assay as a screening test for the diagnosis of deep vein thrombosis in an emergency department. Emerg Med J 21(6):663–666
2. Bramwell R, West H, Salmon P (2006) Health professionals' and service users' interpretation of screening test results: experimental study. Br Med J 333:284–286
3. Henderon LM, Benefield T, Nyante SJ, Marsh MW, Greenwood-Hickman MA, Schroeder BF (2015) Performance of digital screening mammography in a population-based cohort of black and white women. Cancer Causes Control 26(10):1495–1499

Chapter 9
Information

Bayes' Rule shows how we can use evidence to revise belief. Base rates describe what we know before evidence is available. These probabilities are updated once we have the evidence. It is natural to say that we have learned from the evidence. The greater the difference between initial belief, described by the base rates, and the revised belief the more information was provided by the evidence—the more we have learned.

This idea, that we are informed to the extent that we change what we believe, is nothing new. The economist (and statistician) John Maynard Keynes is reported to have said "When my information changes, I alter my conclusions. What do you do, sir?"

So just what is information?

~ • ~

Information is not the same as data. I hear the football results on the radio. My local team when I was a boy, Millwall (currently in the third level of English football, as so often they have been[1]), has won. I am pleasantly surprised. They had been playing Sheffield United, a good team who would go on to win the league, and I thought Millwall would be lucky to draw, even though playing at home, at The Den. Shortly afterwards I meet a friend. "I see your team won", she says. I am not surprised. I already knew they had. Both the radio and my friend brought the same message, the same datum, that Millwall had won. My reactions were different because of what I expected before receiving the message: in the first case surprise, in the second case no surprise at all. The radio broadcast had provided me with information. My friend had not. We are surprised at messages according to how unexpected is the news they bring. Figure 9.1 shows a model.

---

[1]Since writing this Millwall have been promoted to the second level, The Championship, for the coming 2017−2018 season. This sometimes happens.

© Springer International Publishing AG, part of Springer Nature 2018
A. Jessop, *Let the Evidence Speak*, https://doi.org/10.1007/978-3-319-71392-2_9

Fig. 9.1 A model for surprise

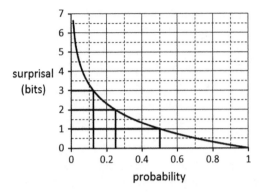

The horizontal axis shows the probability of the message *before it is received*, in other words it shows what I *expect* the message to be. The vertical axis shows how surprised I am when I get the message. Before I heard the radio I thought a win unlikely (low probability) and so was very surprised at the result. After this I knew the result (probability = 1) and so was not surprised at all by what my friend said.

Surprise has no natural units. Call it surprisal and measure it using a ratio scale. Start with a surprisal of 1 when probability is ½. Flipping a coin is like this. The coin is in one of two possible states, heads or tails, and so there are two possible messages to report the result of the flip: "heads" or "tails". Both are equally likely. This unit is called a *bit*, short for binary digit and probably more familiar as a measure of the size of computer memory.

With two coins there are four possible states and messages. The probability of each is halved to ¼ and surprisal increases by one unit to two bits. Every time probability changes by a factor of two (halving or doubling) surprisal changes by a factor of one.[2] This type of scale is quite common in measuring our response to a stimulus: decibels measure our response to sound intensity, changing by one each time sound intensity changes by a factor of ten.

To see how this works look back at the DVT diagnosis of the previous chapter. The base rate showed a prevalence of 27.3% and so before diagnosis the patient would be more surprised to be told he had DVT (probability 0.273) than not (probability 0.727). Figure 9.2 shows this.

---

[2]It is usual to write surprisal $= -\log_2(\text{probability})$. Since surprisal is used only to give relative values other logarithms, to the base 10 or the base e, will do just as well if more convenient. To get the measure in bits just divide by the logarithm of 2, for instance $\log(\text{probability})/\log(2)$ where $\log(.)$ is the logarithm to any base.

Fig. 9.2 DVT base rates

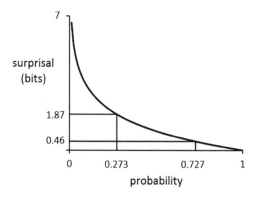

But by definition the more surprising result is the less likely. We can find a measure which takes account of both outcomes by calculating the expected surprisal, the average of the two surprisals weighted by the probability that they happen (Table 9.1)

probability	surprisal	probability × surprisal
0.273	1.87	0.51
0.727	0.46	0.33
1.000		0.85

Table 9.1 Information entropy = expected surprisal

This value, 0.85, is a measure of how surprised you would expect to be, on average, to receive a perfectly accurate diagnosis. It measures what is known and described in base rates before the evidence of the SimpliRed test is collected. This is often called the *information* of the probability distribution. This use of the word information might mislead unless we are alert. What is being measured is the *potential* value of the message (the test result) and this depends on how likely we think each result to be. A better name might be information potential or ignorance. Because of a similarity with a concept in thermodynamics it is also called *entropy*.

Information is independent of the test result. For example, if it was already known that the patient had DVT (probability = 1) there would be no information to be gained by the test. In just the same way if it was known for sure that the patient did not have DVT (probability = 0) there would also be no information to be found from the test. Figure 9.3 shows how information varies with probability. The point shows the information in base rates. The probability that a patient has DVT is 0.273 and so the information entropy is 0.85.

Fig. 9.3 Information entropy and DVT base rate

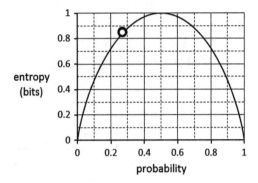

The base rate entropy is the same, of course, whether we choose to plot the probability that the patient has DVT or the probability that there is no DVT (Fig. 9.4).

Fig. 9.4 Both base rate probabilities

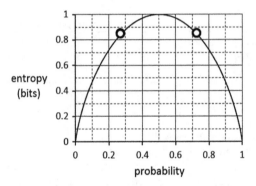

Entropy measures only the *amount* of information, not what the signal tells us.

The two extremes, probability = 1 and probability = 0, describe certainty either that the patient has DVT or does not. Information is zero: given certainty there is nothing left to know. Whichever of these two perfect diagnoses was the result of the test, the information in the base rates is unaffected, though the difference in test results is important for patient and doctor.

The most uncertain base rates, the most ignorant we can be, is when the probability of both events is the same, 0.5. Just as you would expect.

Table 9.2 shows base rate and diagnostic probabilities of the SimpliRed test (from Fig. 8.7). The diagnoses are not perfect but the evidence is useful.

	base rate	SimpliRed test positive	SimpliRed test negative
probability (DVT)	0.273	0.623	0.103
probability (no DVT)	0.727	0.377	0.897
information entropy (bits)	0.85	0.96	0.48

Table 9.2 Information entropies for base rates and test outcomes

A negative test result reinforces the base rate belief that the patient is more likely not to have DVT. The probability of DVT falls from 0.273 to 0.103 and the information entropy also falls, from 0.85 to 0.48 (Fig. 9.5).

Fig. 9.5 Negative result reduces entropy

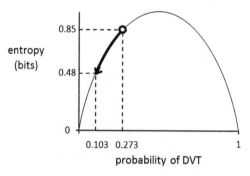

If the test result is positive the situation is different. Now the result has made the presence of DVT more likely. The probability of DVT has increased from 0.273 to 0.623. Not only has there been a switch in the more likely condition but, because of the characteristics of the test, uncertainty has increased, the probabilities are more equal. Information entropy has increased too from 0.85 to 0.96 (Fig. 9.6).

Fig. 9.6 Positive result increases entropy

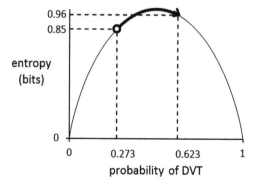

Does this mean that because the test has made us less sure of whether or not the patient has DVT it has not been useful? Well, yes and no. In terms of the information measure uncertainty has increased. But doctor and patient are now alerted that DVT is likely and so further testing would be beneficial, a finding of interest to both. The information measure indicates uncertainty. This has increased; more evidence is now even more valuable.

~ • ~

Information Theory, of which the information measure is a necessary foundation, is due to the mathematician and engineer Claude Shannon (Fig. 9.7). In the 1940s he became concerned with the transmission of telegraph messages which, from our contemporary view, were of limited capacity and error prone.

Fig. 9.7 Claude Shannon (1916–2001)[3]

One person, the source, wants to send a message to another, the destination, but cannot do so directly in face-to-face conversation and so must use some method, some information channel, to do so. The message is first encoded in a form suitable for transmission; natural language for speaking, characters for texting, and so on.

But the channel may be imperfect or noisy so that the signal received is not the same as the signal sent. Figure 9.8 shows Shannon's model of information transmission which is much used.

---

[3]Photograph from Konrad Jacobs. Wikimedia Commons. https://commons.wikimedia.org/wiki/File:ClaudeShannon_MFO3807.jpg. Accessed 19 September 2017.

Fig. 9.8 Information transmission

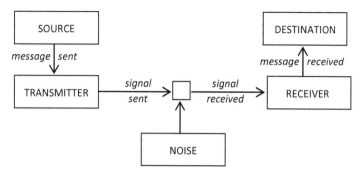

This was not a new problem.

After defeating Napoleon at Waterloo the Duke of Wellington was keen to send a message to London telling of his victory. Among the methods that were available in those days before electrical transmission was semaphore. Flag signals were sent from one signaller to the next in a chain, frequently sited on hill tops, until the destination was reached. The message sent by the jubilant victor was "Wellington defeated Napoleon at Waterloo". But clouds between signal posts meant that only the first two words were sent. Noisy clouds.

Here is a more familiar problem. You and I have arranged to meet and I am to decide where and let you know. We have previously met in a local pub and also sometimes in a newly opened bar called The Hub. I decide that the pub would be better this time. The message I want to send is "See you in the pub". But you are in your car returning from a trip to London so I phone you and say, I like to be brief, "See you in the pub". That is the message. This gets encoded electronically into a signal sent from my phone to yours. The transmission isn't perfect. Your car goes under a bridge just at the time my signal is being transmitted. When you stop for petrol you check your phone and see that you have a message from me. You listen and hear "See ?ou in t?? ?ub." The ? shows where the phone signal is indistinct. The first two missing parts of words present no problem but you are still left with "See you in the ?ub." What does it mean, pub or Hub? You phone me. My phone is switched off. Where have I chosen?

This was the sort of problem in which Shannon was interested. He needed a measure of the effect of the noise in the transmission, the ?s, so that he could assess the benefits of new methods of noise reduction. If you have no other intuition about my intention (equal base rates) you could treat this as a simple problem of making an informed guess about a missing letter, p or h, in a piece of English text. In any language some letters are more common than others. Figure 9.9 shows that in English h appears more frequently than p. The probabilities of h and p are 0.0496 and 0.0205 and so the probability that I said h rather than p is 0.049/(0.049 +0.0205) = 0.706. There is a 71% chance that I said "Hub" and not "pub".

Fig. 9.9 Letter frequencies in English[4]

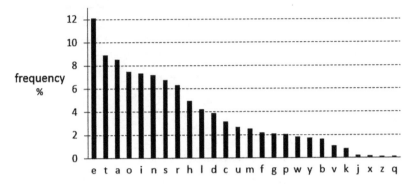

Information entropy is 0.87 (Fig. 9.10), which is high. You decide to go to the Hub but not with much confidence that you have made the right decision. There is a twenty-nine percent chance that when you are at The Hub I'll be waiting for you in the pub.

Fig. 9.10 Information entropy for missing letter

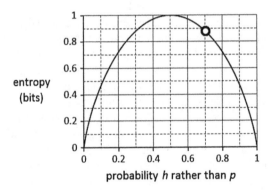

It might be possible to decrease uncertainty and so decrease information (meaning information *potential*, remember) by considering not just a single letter, a monogram, but pairs of letters (a digram) starting with h and p, or trigrams or more.

Or you may have a strong feeling that, for instance, I have a preference for the pub and so may want to incorporate that by using base rates.

~ • ~

---

[4]These letter frequencies are from the website Practical Cryptography. http://practicalcryptography. com/cryptanalysis/letter-frequencies-various-languages/. Accessed 19 September 2017.

Shannon was concerned to devise coding schemes to reduce the problems of noisy communication and so needed some measure of performance. He wanted a measure of how well alternative states or messages could be identified by the receiver. Taking the usual mathematical approach he specified the properties he wished the measure to have and then found the mathematical model which had those properties. The measure which satisfied these requirements was the same as the expected surprisal.[5]

Shannon had his measure but couldn't decide what to call it. To him it was clearly the value of the information obtained by knowing which of a number of possible messages had been sent given that the probability of each is known. A snappier name was suggested to Shannon by the mathematician John von Neumann. When Myron Tribus asked Shannon about his measure,

> Shannon replied: "My greatest concern was what to call it. I thought of calling it 'information', but the word was overly used, so I decided to call it 'uncertainty'. When I discussed it with John von Neumann, he had a better idea. Von Neumann told me, 'You should call it entropy, for two reasons. In the first place your uncertainty function has been used in statistical mechanics under that name, so it already has a name. In the second place, no one knows what entropy really is, so in a debate you will always have the advantage.'" [1]

Other measures have been proposed[6] but the Shannon entropy has such useful properties that it is usually preferred.

~ • ~

Entropy measures only the amount of information, not its meaning. Shannon was not concerned with what you did when you read a message, only that it was transmitted accurately. He was trying to solve the engineering problem, what the receiver did when the signal was decoded was not his concern, though this does not mean that it is a trivial point.

In the early years of the twentieth century traffic lights began to be used in some American cities. One popular design was very simple. A pole was erected in the centre of a crossroads. On the pole was a square box. On each side of the box were two lenses, one red and one green. On two opposite faces the red was above the green and on the other two the green was on top. Inside the box were two electric bulbs, one for the four top lenses and one for the bottom four. The signal was manually controlled so that when a traffic policeman switched on the upper bulb the red lenses on two opposite faces were illuminated and so were the green lenses on the other faces. This signalled the traffic on one road to stop and on the other crossing road to go. Switching on the lower bulb signalled the flowing traffic to stop

---

[5]The now famous formula is $-\Sigma_i p_i \log(p_i)$ where p_i is the ith probability and the sum Σ is over all probabilities.

[6]Most obviously the sum of the squared probabilities. This is known as the Hirshman-Herfindahl index in industrial economics and known to ecologists as the Simpson Diversity Index proposed in Simpson [2].

and the stopped traffic to flow. The system was easy and efficient in transmitting the signal from switch to bulb. The wiring and bulb worked reliably and so there was no noise and signal transmission was perfect. But ten percent of drivers are colourblind. While the signal they received was perfectly transmitted the message they received might be dangerously wrong [3]. The traffic lights used now show the illuminated lens or LED display by both colour and position; red is always at the top. There is some redundancy in using these two properties to show just one signal, but there is a good reason.

~ • ~

Shannon knew nothing of surprisal. I started with that idea because I thought it more intuitively appealing than Shannon's formalism. Shannon published his work in 1948[7] and it caused quite a stir. As well as the engineering implications there were also more naturalistic interpretations, one of which was surprisal [6].

The model of a telegraphic system transmitting information, though imperfectly because of noise, struck a chord particularly with psychologists. They saw a useful analogy with the idea of the human brain as an imperfect transmitter of information about the world—brain as information processor. What Shannon's work offered was a way of measuring information and so a base not just for theory building but for experimentation.[8]

Measuring information transmitted was fundamental and simple enough, it is the difference between what you believe before the message, the base rates, and what you would expect to believe afterwards. To measure what you expect use the average information entropy of each message weighted by how likely each is to occur. The situation for the DVT analysis is shown in Table 9.3.

	base rate	SimpliRed test	
		positive	negative
probability (DVT)	0.273	0.623	0.103
probability (no DVT)	0.727	0.377	0.897
information entropy (bits)	0.85	0.96	0.48
probability	1.00	0.33	0.67
expected information	0.85	0.64	

Table 9.3 Expected information entropy (see Table 9.2)

---

[7]Shannon [4] reprinted with a useful introduction by Warren Weaver as Shannon and Weaver [5].
[8]A short summary of methods and some early results is given by Attneave [7].

Either of the two test results may appear with probabilities 0.67 and 0.33 (Fig. 8.7). The expected information entropy of the test is 0.64 bits. This is smaller than the entropy of the base rates and so something is expected to be gained: information will be transmitted by the test.

The amount of information transmitted is just the difference, 0.21 bits.

Think of this graphically too. Figure 9.11 shows the entropies of the base rates and the two test results

Fig. 9.11 Information entropies for base rates and both test results

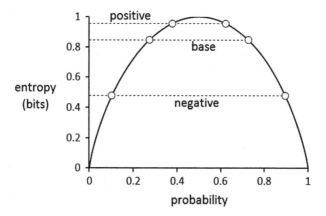

The expected information contained in the test, 0.64 bits, is a reduction of 0.21 bits from the base (Fig. 9.12). This is often called the expected information gain.

Fig. 9.12 Expected information gain

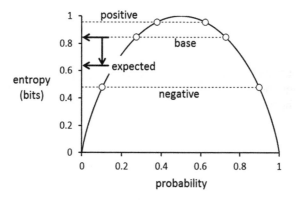

This is a measure of how much information is expected of a test result. Here, as in many cases, the test is set: we use a SimpliRed test. But what if we had to choose between a number of ways of obtaining evidence; different tests, experiments, or whatever? As well as practical considerations of time and money, some more general idea of how much we might expect to learn might be useful too. One of the reasons for collecting evidence is to be better informed, after all. More of this in Chap. 17.

~ • ~

Of the many results of the application of information theoretic ideas by psychologists perhaps the best known is described in a paper by George Miller called *The Magical Number Seven, Plus or Minus Two* [8]. Miller had noted a recurring limit in a number of stimulus-response experiments. Typically, a subject listened to two sounds of different pitch which were given the labels *one* and *two*. Once these labels had been learned the experiment continued with each sound being presented randomly but with equal probability and the subject required to identify it as *one* or *two*. With this small problem, identification was usually perfect. Two equally likely sounds is a one bit stimulus. With four sounds, a two bit stimulus, performance was unaffected. With an eight sound stimulus, three bits, errors were seen. Performance did not improve with more stimuli, of course. In other words, there was a limit to the number of stimuli that could reliably be identified which, averaging the findings over a number of studies of pitch, loudness and so on seemed to be at about seven stimuli, 2.8 bits. Using the information theory model, the number of stimuli is the input, the brain is an information channel, and the number of correct identifications is the output. The limit is the channel capacity. The generalised situation is summarised in Fig. 9.13.

Fig. 9.13 Humans have limited capacity

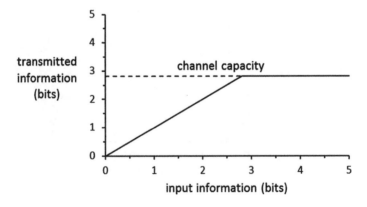

There is much else, of course, for example

musicians with perfect pitch have a much higher channel capacity

capacity can be increased by adding dimensions to the stimuli, both shape and position on a screen for example (remember those traffic lights)

larger problems can be broken into smaller chunks which are within our capacity, a process we would now call modularity

The work was supported by a grant from the Office of Naval Research whose interest in the efficient display of information to pilots and others is obvious.

Miller also noted that the rating scales with which we have become increasingly familiar (on a continuum of agree/disagree, like/dislike) are often no more than seven point scales.

The point to emphasise here is that it was the ability to measure information which made Miller's synthesis possible and introduced the idea of channel capacity to the study of mental processes.

~ • ~

By using evidence we hope to become better informed. Information theory provides a framework for thinking about this process. Entropy is a way of keeping score, of measuring how ignorant we are and so how much there is still to learn, if only we can find the evidence. With the simple problems used in this chapter it is easy enough just to look at the probabilities to see directly how uncertain we are about choosing between alternatives. For larger problems some summary measure is helpful. Entropy is just such a measure.

The signal encoding and transmission model has been influential partly for quantitative modelling, as with Miller, but also as a metaphor for how the mind works. But perhaps this is too big a claim and, more modestly, we should be pleased if the Bayesian point of view offers a description of the results of human reasoning rather than a description of the reasoning process itself. This is a sometime hotly disputed point to which we'll return in Chap. 17.

These ideas about information give a context for considering Bayes' Rule as a process of learning, of using evidence to update an initial belief. It is to the character of these initial beliefs which we now turn.

~~~ ••• ~~~

References

1. Tribus M, McIrvine EC (1971) Energy and information. Sci Am 225(3):179–188
2. Simpson EH (1949) Measurement of diversity. Nature 163:688

3. Petroski H (2016) Traffic signals, dilemma zones, and red-light cameras. Am Sci 104(3):150–153
4. Shannon CE (1948) A mathematical theory of communication. Bell Syst Tech J 27(July/October):379–423; 623–656
5. Shannon CE, Weaver W (1949) The mathematical theory of communication. University of Illinois Press, Urbana, IL
6. Samson EW (1953) Fundamental natural concepts of information theory. *ETC* 10(4):283–297
7. Attneave F (1967) Applications of information theory to psychology: a summary of basic concepts, Methods and Results. Holt, Rinehart and Winston, New York
8. Miller GA (1956) The magical number seven, plus or minus two: some limits on our capacity for processing information. Psychol Rev 63(2):81–97

Chapter 10
Being Careful About Base Rates

We have seen that using a test for medical diagnosis requires that we distinguish two quite distinct sources of information: the characteristics of the test and the characteristics of the patient. Test characteristics are measured by *likelihoods* found by testing and experiment. Patient characteristics are found from data which describe the prevalence of a condition in the population and described as *base rates*. These are brought together in Bayes' Rule

belief is proportional to base rate × likelihood

The calculation was easy. Base rates were found from patient data. A great many cases are just like this.

But what if a doctor wanted to apply the SimpliRED test in a population in which the proportion of patients with DVT was unknown? She wants to know how well the test would predict DVT and so needs base rates. She could in theory make a survey to get some data but there is neither time nor money to do that. She has a number of options. Here are three

ignorance means that base rates can't be given so just use likelihoods
find another similar population and use those base rates
make a judgement based on professional experience

The third option is explicitly based on judgement, but so are the other two. Just using likelihoods means that each is equally weighted, implying that half of the population has DVT. However improbable, this is the judgement implied when base rates are not given. The second option, using base rates from a different population, means that the two populations are thought sufficiently alike to make the analogy worthwhile; another judgement.

Bayes' Rule enables the revision of initial judgement (the base rates) based on data (the evidence). It is to emphasise this judgemental aspect that the revised probabilities in Bayes Grid are labelled *belief*.

The idea that judgement plays a part in the interpretation of evidence is hardly controversial, but for some people expressing that judgement as base rates certainly is controversial. Why is this?

~ • ~

© Springer International Publishing AG, part of Springer Nature 2018   97
A. Jessop, *Let the Evidence Speak*, https://doi.org/10.1007/978-3-319-71392-2_10

It is an oddity of probability that while there is agreement about how to make calculations with probabilities there exists no such universal agreement about what probability is and therefore what legitimately may be described using it. Here are three snapshots from the development of thinking about probability.

In the seventeenth century the French philosopher Blaise Pascal gave an argument aimed at advising agnostics whether or not they should act as if God exists. (The *act as if* is bracingly pragmatic.) The eternal outcomes, heaven or hell, must be weighed against how life is lived, sober rectitude or the enjoyment of more earthly pleasures. The decision depends on how good or bad these outcomes are thought to be but also, critically, on whether or not God is believed to exist. For the believer the probability that God exists is a hundred percent and for an atheist it is zero. Both are certain. Their decisions are easy, though living the life may not be. For an agnostic this probability has some intermediate value depending on the strength of doubt felt. Probability is a measure of belief.

This argument was published posthumously in 1670 and is known as Pascal's Wager [1], the earliest example of what we now call decision theory. Both probability and an assessment of outcomes were important in the decision. In language more familiar to readers of the *Financial Times*, "Pascal wagered God existed because fruitless belief was a better downside risk than damnation."[1]

Time passes. It is the nineteenth century and the dominant mode of scientific argument is experiment. To find whether one strain of rice is more productive than another just plant both and measure the difference. To find the intentions of voters, ask them. Importantly for us, it is natural that the proportion found having a particular value is also the best estimate of the probability of any one rice field or voter has that value. If, in a sample, thirty percent of voters have said they will vote for a particular party then the probability that any one of them will vote that way is also thirty percent. In this view probability is the limiting value of a survey proportion. There is no room here for belief: data are all. The only justifiable estimate of probability comes from the relative frequency with which results occur in an experiment or survey. Those holding this view are frequently called frequentists.

In some important problems there are no data, or not many—R&D spending or most jury decisions, for instance—and yet decisions must be made. Recognising this, methods and models were developed which made use of the probabilities of chancy events and allowed some of these probabilities to be based on judgement. The shift was very much towards providing a formal decision aid for those having to make a decision. The title of Robert Schlaifer's 1959 book—*Probability and Statistics for Business Decisions, An introduction to managerial economics under uncertainty*—shows the intent. Schlaifer writes that "probability will depend on "judgement" in the sense that two reasonable men may well assign different values" [2]. Bayes' Rule fitted into this way of thinking with base rates found from either data or judgement as needed.

---

[1]Jonathan Guthrie in his Lombard column. Financial Times. 7 October 2016 p 24.

And so we come full circle, from theology to management, from belief to judgement. The name changes with context, but the problem does not. The life story of probability can be told in a number of ways [3], but this sketch will do for now.

Probability is a measure of uncertainty; that is common ground. But uncertainty about what? The Bayesian approach is essentially a method for learning, for reducing uncertainty. Start with an initial set of beliefs in the relative truth of competing propositions, alternative models or explanations of what we see. Then evidence modifies those beliefs by means of likelihoods and Bayes' Rule. The main reservation that people have concerns the idea of belief and its use via base rates. And yet it seems quite reasonable that context which cannot be described with data, as prevalence can, should nonetheless be taken into account. For example, you are playing poker and your opponent deals himself a sequence of winning hands. You suspect that he is a cheat rather than just lucky. Would it help to know that your opponent was the Archbishop of Canterbury?[2] The legitimacy of describing judgemental information of this sort using probabilities lies at the heart of the difference between frequentists and Bayesians.

~ • ~

Bayes' Rule rests on two components: likelihoods based on data and base rates which *may* be based on judgement. It is important to emphasise that Bayes' Rule does not depend on judgemental base rates. None of the previous chapters have done so. But because Bayes' Rule has the flexibility to accommodate both judgement and data it has sometimes been presented as synonymous with this subjective approach. This is not so.

It is perfectly possible, as we have seen, to use Bayesian methods without relying on subjective or personal probabilities. We may choose to use base rates based on data. And yet if Bayes' Rule is about updating belief it does seem fairly obvious that these beliefs are *our* beliefs—yours or mine—and necessarily personal. We may both decide to use the same data for our base rates, as the doctors did, and so have the same base rates, but that is more a point of practice than philosophy. So why are some people sceptical about judgemental base rates?

Two of the grounds for this scepticism are, first, that subjective base rates may be chosen to ensure a desired result (cheating) and, second, that in any case humans are poor at making the necessary judgements.

Cheating is nothing new in the presentation of apparently valid analysis of data. Here are two well-known examples.

Sir Cyril Burt, an educational psychologist, claimed that IQ testing was a legitimate way of deciding the educational futures of eleven year old children. The justification was that IQ was largely inherited and so fixed. This claim rested on a study of twins who had been separated at birth and raised apart from each other.

---

[2]This example is usually attributed to Keynes [4].

Some critics were not persuaded. Among other things Burt claimed to have had two female assistants, Margaret Howard and J. Conway, but these have never been found and were almost certainly invented. It seems that he had invented the data too. He was a fraud. His reputation has never recovered.

In 1982 two McKinsey management consultants, Tom Peters and Bob Waterman, published a book *In Search of Excellence*. Their claim was that they had identified excellent companies and the characteristics that made them excellent. The book was a best seller. There was always some scepticism about their methodology and when a number of these excellent companies fairly quickly ran into trouble this scepticism grew. Eventually, in 2001, Peters admitted "This is pretty small beer, but for what it's worth, okay, I confess: We faked the data" [5]. Unlike Burt, this appears to have had no effect on the reputations and careers of Peters or Waterman, which you may think curious and perhaps indicative of the needs of their respective audiences—psychologists and managers.

While neither case involved the subjective assessment of probabilities, you can see why some people are a little careful about accepting the claims of experts.

~ • ~

Cheating is rare and is (usually?) discovered eventually. But what about that second reservation, that we aren't much good at thinking about probabilities? This is a more substantial and much studied claim. But before we move on, here is a question

> Steve lives in the USA and has been chosen at random from a list of registered voters. Here is the view of a former neighbour: "Steve is very shy and withdrawn, invariably helpful, but with little interest in people, or in the world of reality. A meek and tidy soul with a need for order and structure, and a passion for detail." Steve works as one of the following: a baker, a salesperson in a store, a commercial pilot, a librarian, a pharmacist. What do you think Steve does for a living?

Don't read on until you have answered the question.

~ • ~

This problem is a version of one posed by Amos Tversky and Daniel Kahneman in a famous paper [6] which looked at the way we make judgements and the problems that can arise when we try to use probabilities.

What did you decide? The most common response is that Steve is most likely a librarian because the characteristics described by the former neighbour are thought to be stereotypical of librarians. Pharmacists too, perhaps. The point is that although you may or may not have picked librarian as Steve's job your thought process is likely to have been similar: look at the description of Steve and decide which occupation best fits. What you were very unlikely to have done is to think about the information contained in the first sentence.

The way in which Steve was chosen is independent of the characteristics as seen by the neighbour. Even with no knowledge of the data it should be clear that there are a great many more people in retail sales than work as librarians. And yet this is usually ignored. We like the personal, the story-telling, and pay attention to that rather than all that boring stuff about sampling. We latch on to the personal description and see that this fits our preconception of librarians more than others. Tversky and Kahneman called this the *representativeness* heuristic, one of many they identified. Reasoning in this way leads to wrong, ill-justified decisions.

The US Bureau of Labor Statistics gives data for employment at May 2014 (Table 10.1).

Retail salesperson	4,652,160	88
Pharmacist	290,780	5
Baker	173,730	3
Librarian	133,150	3
Commercial pilot	38,170	1
		100%

Table 10.1 Employment by group in US

Whatever you think of librarians this is where you should start. Without knowing anything about Steve's character there is an eighty-eight percent chance that Steve works in retail and only a 3% chance that Steve is a librarian. The odds are almost 30:1. These are the base rates.

Using the Bayes Grid should help (Fig. 10.1).

Fig. 10.1 Partial Bayes Grid for salesperson and librarian only

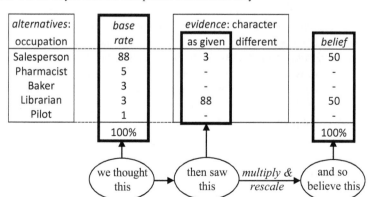

To believe that Steve is a librarian you would have to believe that Steve's characteristics as described by the neighbour are at least thirty times more common among librarians than among salespeople. Perhaps you do, but see how some Bayes thinking helps us ask the right questions (about base rates in this case). Without any data about the different personal characteristics of librarians, bakers and the rest we can go no further.

(In the original Steve was "he"; Steve short for Steven. Or, now, Stephanie? Would it have altered your decision?)

~ • ~

We are none of us immune from these biases.

If you were asked to give as a base rate probability, the proportion of the population with a particular illness, how would you do this without access to data? You would probably use some mix of recollection and analogy. If you or someone close to you had this illness, or if this illness was prominent in news or social media, there is a good chance that you would think it more prevalent in the population than it really is. This is called the *availability* heuristic.

~ • ~

I have in front of me today's newspaper and am leafing through the advertisements

   a jigsaw puzzle for £14.99
   a pair of leather deck shoes for £29.99
   a Zara quilted-pocket rucksack for £45.99
   a tube of Kelo-Cote Scar gel for £28.99
   a book called *Ballerina Body* for £14.99

and so on. What's going on here? Don't we all know that the jigsaw costs £15 and the rucksack £46 and that the penny difference means nothing? Well, no, we probably don't. When we look at the price of £14.99 the first thing that we see, the first thing that registers, is the 14. Then we see the 99 and our perception of the price increases from £14, but almost certainly not enough to conclude that the price is, for all practical purposes, £15. This two stage process is called the *anchoring and adjustment* heuristic. In this case £14 is the anchor value from which an (inadequate) adjustment is made. The result is a biased judgement.

This effect is quite general. For example, the answers given to a question in a questionnaire can depend on whether the previous question required an answer which was a big number or a small number.

These difficulties in the way we selectively retrieve information was just what Tversky and Kahneman were bothered about.

~ • ~

Tversky and Kahneman's paper gave rise to a huge body of work[3] about the way people actually reason rather than how rational models, Bayes' Rule among them, recommend they should. Their point was not just that in reasoning about problems involving some uncertainty we might get the wrong answer, sometimes a probability too big and sometimes too small, but that we are likely to make *biased* estimates and so *consistently* find probabilities which are either too big or too small. Bias is systematic and so you may think that it can be eliminated, by training for instance. This may be optimistic.

Economists realised that this was a useful idea and that economies were made of people and their non-optimal decisions rather than the perfectly informed utility maximisers of much economic theory. And so behavioural economics was born with due recognition paid to Tversky and Kahneman, both psychologists. Tversky died in 1996. In 2002 Kahneman shared the Nobel Prize in economics.

The Tversky and Kahneman heuristics have been influential in the elicitation of probability distributions for Bayes base rates and more generally, though some have proposed alternative explanations for our undoubtedly imperfect performance [9].

~ • ~

Having raised these reasons to be careful let's have a look at the sorts of methods used to get probability assessments from judgements.

Expressing judgements as probability distributions has a fairly long history and, now, a good set of practices for elicitation [10, 11]. Most often it is the opinion of subject experts which is used, or is written up in papers, anyway. This leaves open the question of who counts as an expert and this is usually taken to be someone knowledgeable in the field, such as an ecologist or a forensic scientist. But just because someone has subject expertise does not imply they have expertise in making probabilistic judgements. In some cases, a trial jury is an obvious example, those needing to make a judgement, the jurors, may not even claim subject specific expertise. Whatever the qualifications and experience of the person whose judgement is sought it seems best to be cautious.

But what questions to ask? For the doctor having to decide the prevalence of DVT only one judgement is needed: the percentage of the population with the condition. In cases where more probabilities are needed a scoring system can be used. For example, first rank the alternatives from most likely to least likely. Then give a score of 100 to the most likely and judge a score for the next most likely. Repeat for all alternatives. Finally, rescale to give a base rate probability distribution.

---

[3]There is much to read. Kahneman [7] and Kahneman et al. [8] are two good places to start.

This direct method is not always what you want. Suppose that we need a judgemental base rate distribution for a continuous variable, one that can take any one of an infinite number of values, such as the voting intentions in Scotland. Clearly it makes no sense to ask for lots of individual probabilities such as "what do you think is the probability that 43.1% will vote for independence?" and then "what do you think is the probability that 43.2% will vote for independence?" and so on. But it does make sense to ask questions such as "what is the probability that no more than 40% will vote for independence?". This is a cumulative probability, the probability that the vote is not more than a given value.

People answering this sort of question are also likely to agree that their base rate distribution should follow a smooth curve, one which can be fitted to a set of judgements. The result will not in general perfectly reflect the judgements given because there will have been a smoothing in the fitting process and, in any case, judgements are by their nature likely to be inconsistent to some extent. Knowing where the modelled result and the judgement seem most out of line can indicate this inconsistency and raise the possibility of revision.

Elicitation is used not just for base rates in a Bayes model but any time there is uncertainty about the inputs to a calculation, a profit forecast for instance, or some other sort of risk analysis [12, 13]. The methods used are the same. Here are two examples.

~ • ~

Governments are concerned about the risk of outbreaks of disease in animals. The effects on farmers and on food supply can be severe, whole herds of livestock may be killed and burned to prevent the spread of disease. Following the 2001 outbreak of foot-and-mouth disease in the UK, the Department for Environment, Food and Rural Affairs (Defra) set up a consultation exercise to help decide a better way of managing future outbreaks. Part of this was a study to assess the risk of those outbreaks. A group of veterinarians and economists nominated by Defra were brought together in a series of elicitation workshops [14]. In the first workshop there was a general discussion of the problem and of the idea of elicitation. While some judgements are made by individuals, in a great many cases where responsibility is shared a consensus view can often be developed. The workshops aided that process.

For each of eight specific diseases an estimate of the likely cost was made based on of the cost and frequency of major and minor outbreaks of the disease. The four parameters for which experts' judgement was needed were

the average interval between outbreaks
the probability that an outbreak is a major outbreak
the average cost of a minor outbreak
the average cost of a major outbreak

To illustrate the elicitation we'll look at the judgements made of the probability that an outbreak is a major outbreak for just one disease, classical swine fever, also called hog cholera, a highly contagious disease of pigs.

For each of the four parameters the experts, as a group, had first to decide a value that divided the range of outcomes into two equally likely intervals and then similarly to divide each of those intervals. The result was three values defining four equally likely intervals. This form of elicitation was attractive because deciding the three values required no familiarity with probability, the word need never be used, but the results gave points of cumulative probability: a probability of twenty-five percent that the parameter is less than the lowest value, a probability of fifty percent that the parameter is less than the middle value, and a probability of seventy-five percent that it is less than the highest value.

Kahneman and Tversky's anchoring and adjustment heuristic should give pause for thought. It is likely that the second and third judgements will use the first, the middle of the distribution, as an anchor and so the spread between the two extreme values will be too close, describing a more precise estimate than is intended or justified. In addition, it is well known that experts tend to be overconfident, which also has the effect of compressing the probability distribution.

To counteract any overconfidence, before they gave their three judgements the experts were asked to think about a likely range, extreme values, of the probability distribution. These extremes should have a useful anchoring effect.

Having completed the preliminary thinking about extremes the experts judged the three values to be sixty percent, seventy-five percent and eighty-five percent. These are shown as the three dots on the graph in Fig. 10.2.

Fig. 10.2 Judgements of cumulative probability and best-fit model

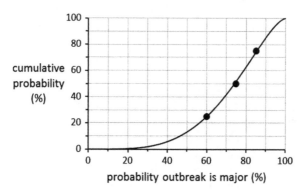

A Beta distribution, just like that used in Chap. 6, was fitted to the three points.[4] The solid line in the graph is found from the Beta distribution. The fit is good. Figure 10.3 shows the Beta distribution in more familiar form

Fig. 10.3 Best-fit model of expert judgement

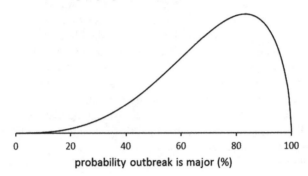

probability outbreak is major (%)

This is the model which best fits the three points and so is satisfactory in that sense. But do the experts agree that it is an adequate description, that it reflects what they think across the range? To test this they were presented with probabilities found from the model, predictions of their judgement. For instance, using this Beta distribution there is a chance of five percent that the probability that an outbreak will be a major outbreak is no more than forty percent. Does this seem right? Using this form of feedback gives the experts the opportunity to revise or augment their judgements. A new model can be fitted.

The main stages of elicitation—setting the context, discussing the problem, making the judgements, fitting a model, checking and revising—are pretty standard but there are, of course, variations in implementation, as the next example shows.

~ • ~

Anthony O'Hagan describes a task he used as part of an introductory session to familiarise subject experts with the idea of elicitation [15]. The task, an everyday sort of problem needing no special expertise, was to estimate the distance between two cities, Birmingham and Newcastle upon Tyne.

The experts were asked to give the extremes, the highest and lowest values for the distance between the two cities. Not only did this fix minds on extremes and so, it was hoped, mitigate the effects of overconfidence, but it also provided a fixed range which was needed to fit the probability model that was to be used (the Beta distribution, again).

---

[4]The Beta distribution has two parameters. Values were found which minimised the sum of the squared differences between the three cumulative probabilities given by the experts and the corresponding values found from the Beta model.

Next, a central value was given. In the Defra case this was the value which divided the range into two equally likely parts. This is called the median and is often used as a measure of average. O'Hagan asked the experts for their most likely estimate, the peak of the probability distribution. This is the mode, another measure of average.

Having fixed the range, between 165 and 250 miles, and the mode, 190 miles, the experts were asked to give five probabilities such as "what is the probability that the distance is less than 205 miles". The distances were chosen to avoid the need for very small probabilities, which are hard to judge. The questions were presented in "a jumbled sequence to avoid problems of anchoring". The Beta curve was then fitted (Fig. 10.4).

Fig. 10.4 Best-fit model of judgements of distance

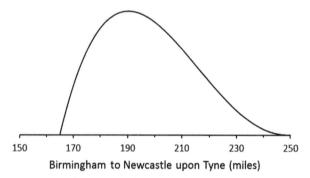

Birmingham to Newcastle upon Tyne (miles)

Judging extreme values, ranges or limits, is one of the more difficult tasks and so O'Hagan used a procedure he called "stepping back". Looking at the fitted distribution he decided whether a wider interval might be justified and, if so, the lower bound is made lower and the higher bound is made higher according to adjustments O'Hagan devised. No other judgements are changed and the Beta model is refitted. The same idea would be used to narrow the range if that seemed as if it might help. Looking at the fitted distribution (Fig. 10.4) he thought the left hand (lower) tail ended rather abruptly and that this might plausibly have been a result of a too narrow range. The stepping back adjustments were made. In Fig. 10.5 the solid line shows the refitted model and the dashed line the original.

Fig. 10.5 Refitted model and original

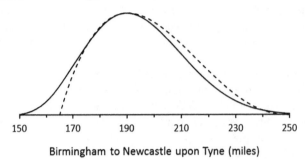

Birmingham to Newcastle upon Tyne (miles)

The result seems to justify O'Hagan's intuition with more probability now located in the left hand tail.[5] The experts accepted this second model of their judgement of the distance between Birmingham and Newcastle upon Tyne.

The true distance is 209 miles.

~ • ~

One of the concerns addressed in these two examples is that when expressing judgements we have a tendency to overconfidence (other biases too, of course). Getting people to think first about extremes uses the anchoring and adjustment heuristic to compensate for this.

A more formal response to the problem was proposed by the physicist Edwin Jaynes [17]. The basic idea is simple: keep the probability distribution as flat as possible consistent with anything you know, or judge that you know. For example, if all you know is the range—from zero to one hundred percent, say—then use a flat distribution with all probabilities equal. If, in addition, you know the average then change the equal probabilities as little as possible to give a skewed distribution with the required average. In other words, keep the distribution as flat as possible consistent with what you know.

As we saw in the previous chapter entropy is at a maximum when probabilities are equal. Jaynes formulated an optimisation problem: choose the probability distribution which maximises entropy subject only to those constraints which describe what you know. This avoids any suspicion of bias.

This is the maximum entropy, maxEnt, methodology [18]. Think of maxEnt distributions as minimally biased or minimally informative. maxEnt methods are used to derive probability distributions useful in themselves and also sometimes as part of Bayesian analysis. The motivating idea—keep base rate probability distributions as flat as possible—is worth bearing in mind whether or not you use the maxEnt mathematics.

~ • ~

---

[5]Not all users of O'Hagan's method have felt this to be necessary, for example see MacDonald et al. [16].

We have assumed that judgements are best made by experts, by which we mean experts in their own fields such as the Defra veterinarians. Subject experts are fairly easy to find, by their jobs usually. But their expertise may not extend to thinking about uncertainty and probability.

Sometimes it is possible to evaluate directly how good an expert is at making probability judgements. For example, in some countries weather forecasters make probabilistic forecasts, such as that there is a sixty percent chance of rain tomorrow. If the forecaster is good at this (well calibrated) it should be the case that the track record shows that on sixty percent of the occasions this forecast was made it did indeed rain. This calibration is possible only if forecasts are made sufficiently frequently to establish a track record, as is the case with some weather forecasters [19], investors [20] and others.

But these circumstances are rare. Most subject experts are not asked to give probabilistic judgements, certainly not sufficiently often to provide a track record or to encourage improvement via feedback.

And, in any case, judgements have to be made by people other than subject experts. A trial may hear from experts but the decision is made by a jury who, as we saw in Chap. 2, are no better at thinking about probabilities than the rest of us. Elicitation is not just for experts.

~ • ~

Four-term U.S. Senator Daniel Patrick Moynihan observed that "Everyone is entitled to his own opinion, but not his own facts". It should be obvious by now how Bayes' Rule might have helped the Senator. In this chapter we have seen some of the problems that make being explicit about judgement difficult and how by being careful about the questions asked it is nevertheless possible to elicit useful probability estimates. To ask in what sense they are true estimates begs a few difficult questions, and so we say useful: those whose judgements are elicited agree that a fair representation has been made of their view and we, who are doing the eliciting, have tried our very best to avoid the effects of bias.

But some people are still a bit nervous about expressing their judgements as base rate probabilities. There are two points to make. First, when evidence is highly discriminatory between alternatives, survey data based on a large sample, for instance, differences in base rates can be swamped by the evidence. Even people with quite different views, and so quite different judgemental base rates, will be driven to practically the same beliefs by the evidence, as we shall see in Chap. 14. This would have reassured the Senator.

Second, it is usually easier for us to react to a numerical estimate rather than provide the number ourselves. When thinking about whether Steve was a librarian or worked in sales it was almost certainly easier to think about whether Steve's characteristics were or were not thirty times more common among librarians than to give an initial judgement. In Chap. 13 we'll see how two statisticians used uniform base rates as a way of not expressing an opinion on a matter unfamiliar to them, but then found it easy to make a decision once they saw the results of their analysis.

There may, of course, be times when the evidence is more equivocal so that your judgement is correspondingly more important. That is why there are arguments.

~~~ ••• ~~~

References

1. Rescher N (1985) Pascal's Wager: a study of practical reasoning in philosophical theology. University of Notre Dame Press, Notre Dame, IN
2. Schlaifer R (1959) Probability and statistics for business decisions: an introduction to managerial economics under uncertainty. McGraw-Hill, New York, p 19
3. Hacking I (1975) The emergence of probability. Cambridge University Press, Cambridge
4. Finklestein MO, Levin B (1990) Statistics for lawyers. Springer, New York, p 93
5. Peters T (2001) Tom peters' true confessions. Fast Company (December). https://www.fastcompany.com/44077/tom-peterss-true-confessions Accessed 21 Sept 2017
6. Tversky A, Kahneman D (1974) Judgement under uncertainty: heuristics and biases. Science 185(4177):1124–1131
7. Kahneman D (2011) Thinking fast and slow. Allen Lane, London
8. Kahneman D, Slovic P, Tversky A (eds) (1982) Judgement under uncertainty: heuristics and biases. Cambridge University Press, New York
9. Kynn M (2008) The 'heuristics and biases' bias in expert elicitation. J R Stat Soc Ser A 171 (1):239–264
10. Garthwaite PH, Kadane JB, O'Hagan A (2005) Statistical methods for eliciting probability distributions. J Am Stat Assoc 100(470):680–701
11. O'Hagan A, Buck CE, Daneshkhah A, Eiser JR, Garthwaite PH, Jenkinson DJ, Oakley JE, Rakow T (2006) Uncertain judgements: eliciting expert's probabilities. Wiley, Chichester
12. Bedford T, Cooke R (2001) Probabilistic risk analysis: foundations and methods. Cambridge University Press, Cambridge
13. Morgan MG, Henrion M (1990) Uncertainty: a guide to dealing with uncertainty in quantitative risk and policy analysis. Cambridge University Press, Cambridge
14. Gosling JP, Hart A, Mouat DC, Sabirovic M, Scanlan S, Simmons A (2012) Quantifying experts' uncertainty about the future cost of exotic diseases. Risk Anal 32(5):881–893
15. O'Hagan A (1998) Eliciting expert beliefs in substantial applications. Statistician 47(1):21–35
16. MacDonald JA, Small MJ, Morgan MG (2008) Explosion probability of unexploded ordnance: expert beliefs. Risk Anal 28(4):825–841
17. Jaynes ET (1957) Information theory and statistical mechanics. Phys Rev 106(4):620–630
18. Jaynes ET (2003) Probability theory: the logic of science. Cambridge University Press, Cambridge
19. Murphy AH, Winkler RL (1984) Probability forecasting in meteorology. J Am Stat Assoc 79 (387):489–500
20. Budescu DV, Du N (2007) Coherence and consistency of investors' probability judgements. Manag Sci 53(11):1731–1744

Chapter 11
Independence

The idea of statistical independence has cropped up once or twice already. It is such an important idea that it deserves a chapter for itself.

In Chap. 6 we saw that by assuming that the voting intentions of our two Scots, Alex and Nicola, were statistically independent it was easy to calculate their joint response by simple multiplication—like flipping two coins. This is a great modelling convenience and an important assumption in much statistical analysis. The calculations are so much easier. Without independence it would have been necessary to describe the relations between the voting intentions of Alex and Nicola and between all pairs of voters. This would be difficult. Independence means that knowing how Alex intended to vote gave us no information about what Nicola intended to do. We didn't have to worry about the relation between them because we assumed there was none. Independence was a good thing.

But not always. In Chap. 8 the hospital doctors in Merthyr Tydfil used the SimpliRed test to see if a patient might have DVT. Diagnostic tests only work if there is a relation between the test result and the patient's medical condition—are they ill or not. If the result of the test is independent of the patient's condition then the test can provide no useful diagnostic information. Independence would be a bad thing.

Being aware of independence and its likely effect is an important part of thinking statistically and of thinking with Bayes. Many times we find neither perfect independence nor its opposite, perfect correlation, but understanding these two extremes and how they might occur is a good place to start. Since this issue is so prevalent more or less any topic could be used for illustration. Two are used here. First, forecasting, to show what happens when several estimates (sources of evidence) are available for the same task. Second, evidence in a jury trial, to show how two or more apparently different pieces of evidence may be misused.

~ • ~

© Springer International Publishing AG, part of Springer Nature 2018   111
A. Jessop, *Let the Evidence Speak*, https://doi.org/10.1007/978-3-319-71392-2_11

In Chap. 4 we looked at the track record of IMF forecasts of the GDP of Japan. Lots of organisations make such forecasts; governments, banks, research institutes and others. In the UK The Treasury publishes each month *Forecasts for the UK economy: a comparison of independent forecasts* which gives the latest one year ahead forecasts from a number of City institutions, mostly banks, and also non-City consultancies, universities and the like.[1]

All forecasters use pretty much the same data but they make different judgements and use different econometric models. It would be surprising if their forecasts bore no relation to each other and also if they were identical. There will be some correlation and this raises two questions: how do we recognise independence and correlation, and what should we do?

To illustrate, consider forecasts made by two banks, Goldman Sachs and HSBC. As in Chap. 4 we are interested only in whether GDP will increase or decrease in the coming year, not by how much. Specifically, how well do forecasts made at year end predict what happens over the next twelve months. To do this compare the one year ahead forecast with outturn.

Figure 11.1 shows UK GDP and forecasts made for 1998 and the next twelve years.

Fig. 11.1 GDP and two forecasts

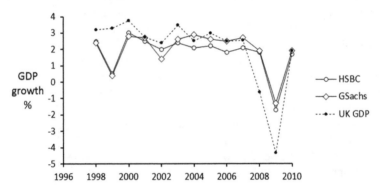

The performance of Goldman Sachs (GSachs) and HSBC look similar. But before we look at their track records let's be clear exactly what is meant by independence and its opposite, perfect correlation.

Each bank forecast five times that GDP would increase and seven times each forecast a decrease. Does that mean that they performed identically and that using either would be as good as using both? Perhaps, perhaps not. What we know of the two forecasters considered separately is shown in Table 11.1.

[1]Current and recent reports can be found at https://www.gov.uk/government/collections/data-forecasts and earlier reports at http://webarchive.nationalarchives.gov.uk/20100407010852/http://www.hm-treasury.gov.uk/data_forecasts_index.htm. Both accessed 22 September 2017.

		Goldman Sachs		
		increase	decrease	
HSBC	increase	?	?	5
	decrease	?	?	7
		5	7	

Table 11.1 Directional forecasts of the two banks

The row totals summarise the twelve HSBC forecasts and the column totals the twelve Goldman Sachs forecasts. But did the two banks make the same directional forecast each year or did they make different forecasts that just happened to be the same in aggregate—five increases and seven decreases? What would the two extremes look like?

In Table 11.2 see what happens if the forecasts that UK GDP would increase or decrease were always the same for both banks.

		Goldman Sachs		
		increase	decrease	
HSBC	increase	5	0	5
	decrease	0	7	7
		5	7	

Table 11.2 Perfectly correlated forecasts

The forecasts are perfectly correlated. Once we know what HSBC forecast we can predict perfectly what Goldman Sachs will forecast, and vice versa. This is what perfect correlation means. The likelihood distributions describe this dependency (Table 11.3).

		Goldman Sachs		
		increase	decrease	
HSBC	increase	100	0	100%
	decrease	0	100	100%
	unknown	42	58	100%

Table 11.3 Likelihoods for perfectly correlated forecasts

With no knowledge of the HSBC forecast the best estimate of what Goldman Sachs will forecast is to use its track record. Five of the twelve forecasts, forty-two percent, predicted an increase. This is the likelihood that Goldman Sachs will forecast an increase. But if we already know what HSBC forecast the likelihood is either a hundred per cent or zero for the Goldman Sachs forecast. The HSBC forecast contains perfect information about the Goldman Sachs forecast (and vice versa, of course).

At the other extreme, if the two forecasts were independent the HSBC forecast would contain no information about the Goldman Sachs forecast. The likelihoods would be the same (Table 11.4).

		Goldman Sachs		
		increase	decrease	
HSBC	increase	42	58	100%
	decrease	42	58	100%
	unknown	42	58	100%

Table 11.4 Likelihoods for independent forecasts

HSBC forecast an increase five times so we would expect forty-two percent of these five forecasts, two in round figures, to be for an increase (Table 11.5).

		Goldman Sachs		
		increase	decrease	
HSBC	increase	2	3	5
	decrease	3	4	7
		5	7	

Table 11.5 Frequencies for independent forecasts

To calculate this distribution of all four joint forecasts in the table just knowing how many times each bank forecast an increase and a decrease is enough. The assumption of independence and simple multiplication does the rest.

~ • ~

What really happened? In the twelve years from 1998 UK GDP increased six times and decreased six times. The track records of the two forecasters are shown in Tables 11.6 and 11.7.

		HSBC		
		increase	decrease	
GDP	increase	5	1	6
	decrease	0	6	6
		5	7	

Table 11.6 Track record of HSBC forecasts

		Goldman Sachs		
		increase	decrease	
GDP	increase	4	2	6
	decrease	1	5	6
		5	7	

Table 11.7 Track record of Goldman Sachs forecasts

These track records are different, though not by much. The two forecasters differ only in two years (Table 11.8).

		Goldman Sachs		
		increase	decrease	
HSBC	increase	4	1	5
	decrease	1	6	7
		5	7	

Table 11.8 Forecasts of both banks

It seems reasonable tentatively to conclude that Goldman Sachs is unlikely to give forecasts which are much different from those of HSBC. But whatever the correlation between forecasts (sources of evidence) how can their joint effect be modelled?

As we have done before, the track records can be shown as likelihoods though noting, with due caution, that the percentages are based on only twelve values. (Tables 11.9 and 11.10).

		HSBC		
		increase	decrease	
GDP	increase	83	17	100%
	decrease	0	100	100%

Table 11.9 Likelihoods for HSBC forecasts

		Goldman Sachs		
		increase	decrease	
GDP	incr ease	67	33	100%
	decrease	17	83	100%

Table 11.10 Likelihoods for Goldman Sachs forecasts

We want a forecast that combines those provided by HSBC and by Goldman Sachs. It seems reasonable to assume their forecasts are not independent and that it would be a mistake to treat them as if they were. But how big a mistake? If both banks forecast a decrease and we assumed independence we would simply multiply the likelihoods. We have no strong views of our own, and so equal base rate probabilities (Fig. 11.2).

Fig. 11.2 Using two forecasts and assuming independence

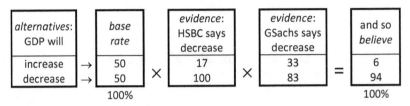

(This is a different and stripped-down version of the Bayes Grid. Only the likelihoods needed for this analysis are shown. You should know all about rescaling by now.)

Believe that there is a ninety-four percent probability that GDP will decrease in the coming year.

Multiplying likelihoods is justified by the assumption of independence, though we are sceptical that this is justifiable. Without independence we need to describe the effects of the correlation somehow. A straightforwardly direct approach in this case is to look at all four possible combinations of the two forecasts (Table 11.11).

HSBC & Goldman Sachs

		↑↑	↑↓	↓↑	↓↓	
GDP	increase ↑	4	1	0	1	6
	decrease ↓	0	0	1	5	6
		4	1	1	6	

Table 11.11 All four combinations of forecasts by the two banks

On the four occasions when both banks predicted an increase there was an increase. On five of the six occasions when both forecast a decrease there was a decrease. Likelihoods for all four possibilities are shown in Table 11.12.

HSBC & Goldman Sachs

		↑↑	↑↓	↓↑	↓↓	
GDP	increase ↑	67	17	0	17	100%
	decrease ↓	0	0	17	83	100%

Table 11.12 Likelihoods for combined forecasts

If both HSBC and Goldman Sachs forecast a decrease we should believe with probability eighty-three percent that that is what will happen (Fig. 11.3).

Fig. 11.3 Both banks forecast that GDP will decrease

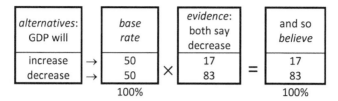

The assumption of independence, however convenient, gives an unjustifiably high probability that GDP will decrease; ninety-four percent rather than eighty-three percent. This is general: unjustified assumptions of independence lead to an overvaluation of the evidence.

Treating correlation problems is typically quite a bit more difficult than this, but these calculations work for small problems. This illustration is not meant to be a primer on forecasting but to make clear why correlated sources might be problematic and to give an idea of the issues involved in assessing their effect.

~ • ~

Independence makes analysis much easier, though it may sometimes be that analytical convenience wins over what can be justified by the data. For this reason Bayes methods which make this assumption are sometimes called naïve Bayes methods. The assumption can always be checked by comparing what we see in the data with what we would expect to see given the assumption of independence, provided that the track record is available.[2]

Without independence the interactions have to be explicitly estimated and this is likely to need more data. The illustration was based on twelve annual forecasts, not many but possibly hard to increase. Taking more years is open to the objection that circumstances and models have changed so much that an extended time span provides only a false guide. On the other hand, taking more data within the time span—quarterly, for example—presents its own problems. The forecasters may not give forecasts that frequently, or perhaps at different times.

Having so few data means that estimating the frequency of rare events is particularly difficult. Although it is certainly true that in none of the six years when GDP decreased did HSBC forecast an increase can we be sure that it never would? Spreading the data over eight categories rather than four increases the problem. Although this is an illustration it uses real data. This point about increasing the demands on available data is quite general. Estimating the degree to which

---

[2]To compare 2 × 2 or larger tables there is a standard method. Find any statistics book and look for the chi-squared (χ^2) test for contingency tables.

evidence from different sources are correlated and the effects of that correlation is likely to need more data and extended analysis, which itself often involves a few simplifying assumptions.

~ • ~

Wrongly assuming that different sources of evidence are independent can lead to biased results: the evidence will seem to be more discriminatory than it is. The same problem might affect the relation between evidence and base rates.

In medical diagnosis the likelihoods of evidence describe what we know about a test based on experimental results and base rates describe what we know about the patient based on factors other than those measured by the test (age, gender, occupation...) and how this affects the chance of having an illness. Both evidence and base rates are found from data but the data are quite different and so may reasonably be treated as independent.

This is not always so. For the GDP forecasts the likelihoods used with evidence come from the track records of the forecaster. They are data based, like the likelihoods used to assess the evidence provided by a medical test. But there are differences. There are no controlled experiments with an economy (we all hope) and so data on which to base a track record are naturally limited and might not easily be increased. A drug company can increase the sample, an economist will find this much harder.

If the base rates are an expression of judgement where did that judgement come from? Perhaps you are a professional economist and so have been trained in the same way of thinking as the economists responsible for the forecasts. You all probably read the same journals and newspapers and websites. And unless you have been freakishly diligent you will not have kept a track record of your own base rate estimates. This makes finding the correlation between base rates and evidence difficult. A safety first approach is to have no view and so modestly use equal base rates as above. A variation on this is to have these equal non-judgemental base rates and then, after the analysis of the evidence, judge whether you feel so strongly that a different conclusion is for you. If you felt that your base rates might plausibly be greater than eighty-three percent in favour of an increase then decide that rather than the decrease indicated by the evidence of the banks' forecasts alone. That's what we did with Steve (Fig. 10.1) and we'll see another example of this in Chap. 13.

Forecasts based on judgement rather than mathematical models present similar problems. Human judgement is fallible, of course, but if we have a track record it may still be of use, as with Lawro's football forecasts.

~ • ~

Statistical independence, our concern here, is not the same as other sorts of independence. An economic forecaster may be labelled as independent meaning that its funding and policy are independent of the government, but the forecasts provided by both are not necessarily statistically independent, they may be highly correlated, as were the forecasts of our two banks.

The problem is more difficult if evidence is judgemental. In financial markets shared views sometimes drive investment in strange directions, from the tulip mania of the seventeenth century to the dot-com bubble at the end of the last century to, perhaps, current valuations of some Silicon Valley startups.

The wisdom of the crowd? Market sentiment? A reassuring consensus among experts? It might all be nonsense, of course, amplified by the positive feedback of conversation, journalism, websites and the rest. Perhaps just herd instinct, the fear of not being one of the crowd.

It is unlikely that any one of us is immune.

Even with these reservations there is evidence for a simple approach. It was the observation of the statistician Sir Francis Galton that the average of the estimates of the weight of an ox made by people at a country fair in 1906 was within one percent of the true weight. This was the original crowd wisdom.

Taking a simple average is appealing and likely to work well for many problems, particularly forecasting problems.[3] All methods have some bias depending on methodology and data. Taking the average of about five forecasts seems to smooth these biases and give a better forecast that just one method.

The correlation question is always there.

At the extreme we may hear the same from a number of experts, however defined. Treating them as independent sources will ultimately lead to convergence to near certainty. But if all are just repeating the same received view this certainty may be an illusion. The Bayes method allows for this *provided we have the data to establish track record*, which we may not.

Suppose some economic forecaster always produced exactly the same forecasts as Goldman Sachs. Perhaps Goldman Sachs is held in such high esteem that others simply repeat its forecast. Or they may all, Goldman Sachs included, be very influenced by the World Bank. As a result it is never the case that one forecasts an increase and one a decrease. This other forecaster adds nothing to what Goldman Sachs tells us. It is a clone. The likelihoods will show this (Table 11.13).

Goldman Sachs & clone

		↑↑	↑↓	↓↑	↓↓	
GDP	increase ↑	67	0	0	33	100%
	decrease ↓	17	0	0	83	100%

Table 11.13 Likelihoods for Goldman Sachs and clone (see Table 11.10)

Third or fourth or fifth forecasters would also add nothing if they too are clones. The difference between a possibly false convergence and a number of forecasts which are effectively clones can be large after only a few forecasts. Here is what happens. Based only Goldman Sachs' forecast and track record believe that there is an eighty percent probability that GDP will increase (Fig. 11.4).

---

[3]There are a number of surveys. Armstrong [1] is a good place to start.

Fig. 11.4 Goldman Sachs forecasts GDP increase

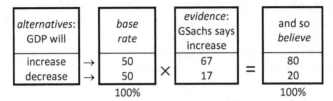

If all other forecasts are from clones and we treat them as such (perfect correlation) the forecast will not change. If, on the other hand, a second forecast from a clone is treated as independent there will be a change (Fig. 11.5).

Fig. 11.5 Goldman Sachs and clone both forecast GDP increase

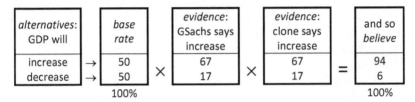

The probability of an increase rises sharply. More such forecasts only make it seem all but inevitable that GDP will rise. But if the other forecasters are clones what they say has no effect (Fig. 11.6).

Fig. 11.6 Forecasts by clones and by independent forecasters

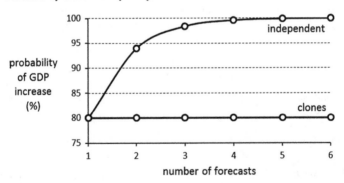

Identical, or very similar, forecasts may be due to this clone-like behaviour or the same coincidence might indicate consensus. Forecasts made using different methods might give very similar results because the system (the economy in this case) is so constrained that its trajectory is largely predetermined, at least in the near future. Look at the forecasts of HSBC and Goldman Sachs. In this case treating the forecasts more like independent estimates than clones might be wise. Not a particularly easy judgement. Investigate further.

~ • ~

Failure to recognise independence, or to see it when it isn't there, can have serious consequences. Perhaps the most worrying possibility is in the conduct of jury trials.

Juanita Brooks lived in the San Pedro district of Los Angeles. On a June day in 1964 she was robbed of her purse. Based on descriptions provided by Ms. Brooks and John Bass, an eye witness, the police arrested Janet and Malcolm Collins. In court the prosecution produced as evidence the probability that any couple, chosen at random, possessed the characteristics given by Brooks and Bass (Table 11.14). Janet and Malcolm Collins did.

Characteristic	probability
Partly yellow automobile	1 in 10
Man with moustache	1 in 4
Girl with ponytail	1 in 10
Girl with blond hair	1 in 3
Negro man with beard	1 in 10
Interracial couple in car	1 in 1,000

Table 11.14 Probabilities of robbers' characteristics given by the prosecution

The prosecutor, Ray Sinetar, called Daniel Martinez of the nearby California State University at Long Beach. Given these probabilities *and that the characteristics were independent* Martinez correctly testified that by simple multiplication it followed that there was a probability of just 1 in 12 million of these six characteristics being seen together. Sinetar presented this as strong evidence of guilt. Janet and Malcolm Collins were convicted.

This case is infamous for a number of reasons [2, 3], not least that the six probabilities were given by Sinetar as likely typical values. In other words, he guessed. (Did a quarter of all men wear a moustache? It was the sixties and it was California, but still...). What we need to think about here is the assumption of independence, also proposed by Sinetar. Most obviously, the probability that a man with a beard also has a moustache is surely higher than the probability that a man with no beard has a moustache. Taking Sinetar's estimates at face value means that twenty-five percent of men had a moustache and ten percent had a beard. Show these as row and column totals, as we did with the banks (Table 11.15).

	beard	no beard	
moustache	?	?	25
no moustache	?	?	75
	10	90	100%

Table 11.15 Probabilities robber had a beard and robber had a moustache

What if Sinetar was right, that wearing a beard and wearing a moustache are independent? In that case just ten percent of the twenty-five percent of men with a moustache also had a beard, two and a half percent of all men (Table 11.16).

	beard	no beard	
moustache	2.5	22.5	25
no moustache	7.5	67.5	75
	10	90	100%

Table 11.16 Probabilities robber had both a beard and a moustache, assuming independence

On the other hand, it seems plausible to believe that the overwhelming majority of men with a beard also had a moustache. At the extreme assume that *all* bearded men have a moustache, the proportion of bearded men with no moustache is zero (Table 11.17).

	beard	no beard	
moustache	10	15	25
no moustache	0	75	75
	10	90	100%

Table 11.17 Probabilities robber had a beard and robber had a moustache, all bearded men also have a moustache

In this case the probability that a man has a moustache and a beard is ten percent, four times the two and a half percent if independence is assumed. The probability of one in 12 million given by Sinetar is reduced by a quarter to one in three million. Still a low probability but not as low. Even if we admit that a small number of men wore a Dutch style of beard (ear to ear via the chin) with no moustache this is unlikely to change the higher estimate much.

You can probably see other possible dependencies which would increase the probability even more. Janet and Malcolm Collins may not be that extraordinary after all. Identifying them as guilty looks less convincing. The effect of independence, real or assumed, is, once again, to increase the diagnostic power of the evidence.

More on problems with the law in Chap. 16.

~ • ~

Using evidence from a number of sources is easier if these sources are statistically independent, or if they are assumed to be. Just multiply the likelihoods. If the sources are not independent the calculations are more difficult. The convenient assumption simplifies calculation but can lead to quite large errors.

Building a mathematical model, or any sort of model, requires judgement about which simplifications are justified and which are not. This is an important skill. Deciding to assume that two or more sources of evidence are independent even though we know, from theory or from data, that they are not is one such judgement. Sometimes the simplified model is good enough. The results justify the simplification. Here is an example.

Classification is an old problem. The ancient Greeks, and before them the Chinese and the Aztecs, saw the usefulness of grouping similar plants into classes. There are fewer groups than there are plants so the reduction in variety helped to make sense of the world as they saw it. Classification is still useful for this same reason. With the much increased availability of data and of the computing power for analysis all sorts of objects are classified. In one sense, of course, all this book is about classification, about deciding which alternative is the most likely and acting upon it—DVT or not, choose another door or stick, and so on.

Classifying documents is increasingly useful.

Words are everywhere; in tweets and emails, in news stories and hotel reviews, in online tutorials and books. Call any such document a text.

Many people have an interest in using words and patterns of words to decide a classification. For example, librarians find it useful to classify books and other documents as an aid to retrieval in answer to a query. This sort of categorisation is based on what a text *is*. A related, and more difficult, problem is to decide what a text *says*; are your emails indicating a favourable view of a particular product, or perhaps they can be used to find out which way the political wind is blowing [4]. There are many other applications, from market intelligence to national security, details of which are, for obvious reasons, secret.

This problem, deciding feelings rather than topic, is called sentiment analysis. New methods for making these classifications are being developed all the time. Just about the simplest is the Naïve Bayes Classifier. The basic idea is that the relative frequencies with which certain words or groups of words (generically called features) indicate the sentiment being communicated. For a Bayesian analysis these data give likelihoods of feature occurrence for texts conveying alternative sentiments; happy/sad, Democrat/Republican and so on. This requires a set of texts (the training set) which are already classified by human assessors or by some other means. Once this is done the likelihoods can be found and used to classify new texts using the method shown in the Bayes Grid. But how to combine the evidence, the different features, none or many of which may appear in any text? Make the simplifying (hence "Naïve") assumption of independence, that the occurrence of "love" is independent of the occurrence of "delight", that "Trump" is independent of "Republican". This gives a simple model relying only on likelihoods for each feature and then multiplication. The text is put into the most likely category, the category for which the belief probability is highest.

An illustration is provided by Bo Pang and Lillian Lee of Cornell University and Shivakumar Vaithyanathan of IBM [5]. Their task was to categorise movie reviews as either positive or negative. The training set was 752 negative reviews and 1301 positive reviews taken from the archive of the Internet Movie Database (IMDb). After some experimentation these seven words were used as evidence of a positive review

love, wonderful, best, great, superb, still, beautiful

and for a negative review these five words and two symbols

bad, worst, stupid, waste, boring, ?, !

It may not be immediately obvious why some are included, but think about it: "Still, though, it was worth seeing" or "What was the director thinking about?"

Three methods of classification were used with eight versions of the test data. All three methods gave accurate classification in about eighty percent of cases, the differences in accuracy being no more than one or two percentage points. The Naïve Bayes Classifier was easily the simplest of the three and by no means the worst performer.

Naïve Bayes Classifiers are easy to implement and quick to execute. One reason for their success may be that the values of the probabilities for each category are used only to find the largest, sufficient for the categorisation, rather than the values themselves. Correlation between the word pairs may not have much effect on which is the largest probability, whatever its effect on the value.

Because it gives probabilities, which most other classification methods do not, it would be possible to have a threshold value to prevent classification when the maximum probability is low. When sifting very big data sets the need for quick and automatic classification is likely to make this unfeasible, or perhaps commercially undesirable: "not sure" doesn't sound too helpful when targeting some advertising. We have seen that assuming independence increases apparent certainty which reinforces this possibility of overconfident classification.

Notwithstanding these reservations, Naïve Bayes Classifiers have been particularly successful in information retrieval and less so, but still respectably good, in categorisation generally. Their success probably depends on the complexity of the data: the simpler the better [6]. It seems that even when there is strong dependence between word pairs or other attributes this classifier still performs well [7].

~ • ~

I feel professionally obliged to point out, in a phrase much used by stats teachers, that correlation does not equal causation. Just because we see that there is a pattern—that big values of something consistently co-occur with big (or small) values of something else—does not mean that variation in one *causes* variation in the other. The most famous example is probably the apparent relation between storks and babies. Why else all those "welcome to your new baby" cards with the new arrival dangling rather precariously from the large beak of the helpful bird? Here is a graph which seems to support the folk tale (Fig. 11.7).

Fig. 11.7 Do storks bring babies?

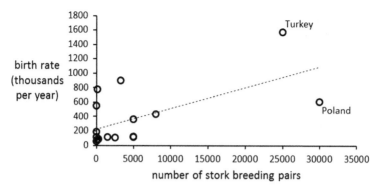

The points show the human birth rate and the number of breeding pairs of storks in sixteen European countries and Turkey [8]. Looks good? But storks cannot be responsible for human births, can they? A quick web search will be entertaining.

~ • ~

Correlated sources are most likely when there is functional dependency or when different organisations or individuals use similar mental or mathematical models and so arrive at similar conclusions. Evaluating the information contained in them all taken together should ideally recognise their joint occurrence. The success of the Naïve Bayes Classifier shows that sometimes assuming independence can give a useful result. This is not a general conclusion; context is all. For classification the assumption can be justified by results. For evidence in a jury trial making the wrong assumption can be a matter of life and death, or something close to it.

Statistical independence, or the lack of it, is important. Bayesian analysis can take account of non-independence, though it may complicate the calculations.

~~~ ••• ~~~

References

1. Armstrong JS (2001) Combining forecasts. In: Principles of forecasting: a handbook for researchers and practitioners. Springer, New York. pp 417–439. Available online at http://repository.upenn.edu/cgi/viewcontent.cgi?article=1005&context=marketing_papers. Accessed 22 Sept 2017
2. Koehler JJ (1997) One in Millions, billions and trillions: lessons from People v. Collins (1968) for People v Simpson (1995). J Legal Educ 47(2):214–223
3. Schneps L, Colmez C (2013) Math on trial: how numbers get used and abused in the courtroom, Chapter 2. Basic Books, New York, pp 23–38
4. Durant K, Smith MD (2007) Predicting the political sentiment of web log posts using supervised machine learning techniques coupled with feature selection. In: Nasraou O, Spiliopoulou M, Srivastava J, Mobasher B, Masand B (eds) Advances in web mining and web usage analysis. Lecture notes in computer science, vol 4811. Springer, Heidelberg, pp 187–206

5. Pang B, Lee L, Vaithyanathan S (2002) Thumbs up? Sentiment classification using machine learning techniques. Proceedings of the 2002 Conference on Empirical Methods in Natural Language Programming (EMNLP 2002). pp 79–86. http://www.cs.cornell.edu/home/llee/papers/sentiment.pdf Accessed 22 September 2017.
6. Lewis DD (1998) Naive (Bayes) at forty: The independence assumption in information retrieval. In: Nédellec C, Rouveirol C (eds) Machine learning: ECML-98. Lecture notes in computer science, vol 1398. Springer, Heidelberg, pp 4–15
7. Domingos P, Pazzani M (1996) Beyond independence: conditions for the optimality of the simple Bayesian classifier. In: Saitta L (ed) Machine learning, proceedings of the thirteenth international conference (ICML '96). Morgan Kaufmann, Burlington, MA, pp 105–112
8. Matthews R (2000) Storks deliver babies (p = 0.0008). Teach Stat 22(2):36–38

Chapter 12
Review

The cases in the first section, the first six chapters, used only likelihoods; no base rates. Now that we know about base rates we need to revisit those cases to see what effect contextual or judgmental inputs might have had. What did we miss by adopting this simple Bayes approach?

That is looking back. Looking forward, the next and final section shows how Bayes thinking has been used to help in a number of different cases. Some of these applications need more complicated models of likelihood than we have seen so far. The principles that are used to structure the problems will be familiar but the mathematics will need more than simple rescaling. Discussing the details of these methods is not my purpose. I want to show how the interplay between alternatives, evidence, likelihoods and base rates can help in difficult situations. For each case you'll have references that will enable you to find out more if you wish. However, you might find it helpful to have just a sketch of the ideas underlying some of these more advanced calculations. This is given at the end of this chapter.

But first, the review of earlier cases. Would base rates have helped?

~ • ~

Our two Swedish policemen, Larsson and Mankell, had Ingrid's identification of the colour of the thief's car as blue and a chart showing the likelihoods that she would say the car was blue (Stig's car) or green (Jan's car). That was all. They used only that evidence. They did not make use of any other information or intuition about who was the more likely to be guilty, a case of equal base rates (Fig. 12.1).

© Springer International Publishing AG, part of Springer Nature 2018   127
A. Jessop, *Let the Evidence Speak*, https://doi.org/10.1007/978-3-319-71392-2_12

Fig. 12.1   Using Ingrid's evidence of car colour

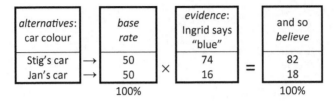

Are these equal base rates always the right way to go? What about the brilliantly intuitive detective we see in all those crime shows? Well, perhaps this is what happens. But probably not. The more likely is that some other evidence is available. For instance, an eyewitness briefly sees the face of the thief as he tears off his ski mask before jumping into the (blue or green) car. Stig and Jan are about the same age and ethnicity so Larsson and Mankell know this is a long shot. In any case, eyewitness identification isn't that reliable [1]. Still, they decide to see if their witness will pick either a photograph of Jan or one of Stig.

From what they know Larsson and Mankell believe that there is not much more than a 50:50 chance that the eyewitness can pick the right man: say sixty percent. The witness could pick Stig even though Jan is guilty, or vice versa, or perhaps neither. The detectives think these identifications about equally likely. Table 12.1 shows the likelihoods.

alternatives:	evidence: eyewitness identifies			
the thief is	Stig	Jan	neither	
Stig	60	20	20	100%
Jan	20	60	20	100%

Table 12.1   Likelihoods for eyewitness identification evidence

The witness is shown the two photographs and asked if either is the man that ran from the shop. The police are in luck. Their witness points to the picture of Stig; "That's him". This is useful evidence (Fig. 12.2).

Fig. 12.2   Using eyewitness identification

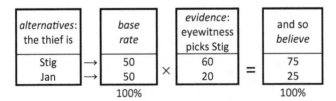

Larsson and Mankell can reasonably believe there is a seventy-five percent chance that Stig is their man. When, next, they hear Ingrid's evidence they use what they have learned from the eyewitness as their base rates (Fig. 12.3).

Fig. 12.3 Using Ingrid's evidence of car colour after the eyewitness identification

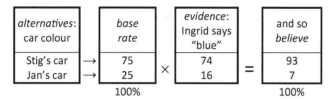

Base rates describe what we believe before evaluating new evidence: they encode judgement and other evidence seen up to that point. Setting out the police investigation as two distinct stages emphasises this.

It can be helpful to think of evidence in this way, following the time line of the police enquiries. But because we consider the different pieces of evidence to be statistically independent the ordering has no effect on the result. From a purely computational point of view all evidence can be considered in one calculation, either with the eyewitness identification first (Fig. 12.4) or with Ingrid's car colour evidence first (Fig. 12.5).

Fig. 12.4 Two pieces of evidence; eyewitness first

Fig. 12.5 Two pieces of evidence; Ingrid first

The order is irrelevant.

The order in which evidence is given in court could also be irrelevant but is more likely to be decided by lawyers so as to present their arguments most forcefully. More on the law in Chap. 16.

~ • ~

What of Lawro's track record and his forecasts?

Of the three alternative match results—away win, home win and draw—Lawro forecasts an away win. We should believe that this is the most likely result, probability about fifty percent (Fig. 12.6).

Fig. 12.6 Using Lawro's prediction

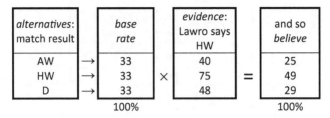

(The base rates sum to 99, I know, but let's choose easy presentation.)

What would be the argument for different base rates? What information might you have? If you are a football fan you'll certainly have a view (judgement may not exactly be the right word) and so will your friends and so will the newspapers, tv shows and the rest you all watch and read. The problem here is independence. Statistical independence, that is.

We looked at this in the previous chapter and saw how to deal with two pieces of evidence which were not independent, provided we had data which enabled the dependency to be described. We would also want evidence to be independent of base rates for the simple multiplication of the two to be justified, and this might be tricky here.

Best stick with equal base rates but enjoy the arguments with your friends.

~ • ~

Chapter 5 showed how thinking about likelihoods could help a game show contestant, Craig, decide which door to open—yellow or blue or green. The whole point of this example was that although it was agreed that at the start of the game all three doors were equally likely to be the way to the prize those equal probabilities should be revised once one of the doors has been opened.

Were those initial base rates justified?

Yes, they were.

Craig may have had a preference for one colour over the others—perhaps blue was his favourite colour—and this influenced his decision but this is irrelevant. He had no reason to believe that the prize was more likely to be behind one door than another. Perhaps there were people who studied the show and thought they could detect a bias, that the prize was most often behind the yellow door, but this is more evidence of obsession (a game show nerd) than expertise. Equal value base rates are justified.

~ • ~

The final case was the opinion survey for the Scottish referendum (Chap. 6). In the YouGov poll forty-three percent of the 864 respondents said "Yes", they would vote for independence. The analysis used only the Binomial likelihoods and the survey data. We now know to interpret this as equal valued base rates, which in this case means a flat uniform distribution (Fig. 12.7).

Fig. 12.7 No initial opinion so use equal base rates (from Fig. 6.9)

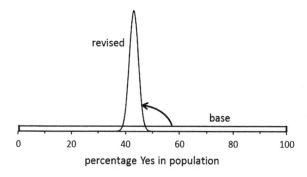

The arrow shows the learning, from the flat base (all values equally likely) to the revised distribution showing what we are justified in believing given the poll result.

But we may have had an opinion, a judgement, of the strength of feeling in favour of independence. We may be pro-independence and believe that support for a Yes vote is over fifty percent, say "around sixty percent". With the methods described in Chap. 10 this judgement can be expressed as a Beta distribution. Using this as our base rate gives a new analysis[1] (Fig. 12.8).

Fig. 12.8 Initial opinion favours the Yes vote

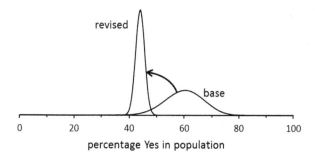

---

[1]Using a Beta distribution in this way is convenient and quite common. The distribution has two parameters, ALPHA and BETA. For the uniform (flat) base distribution ALPHA = 1 and BETA = 1. For the illustrative judgemental distribution, "about 60%", ALPHA = 30 and BETA = 20.

To revise given the survey data add the number of Yes respondents to ALPHA and the number of No respondents to BETA. There were 372 who answered Yes (43% of 864) and 492 No. The revised parameters are

ALPHA = 30 + 372 = 402 and
BETA = 20 + 492 = 512.

The result is not much different from that found using flat base rates. With a flat base distribution the ninety-five percent confidence interval was from 40% to 46%. With the pro-independence judgemental base this changed to 41% to 47%. The estimate is a little higher. This can be seen from comparing the two ranges. The average has increased half a point from 43.5% to 44%. This is to be expected. Our judgement was "about 60%" and combining that with the YouGov result, "about 43%", has increased the estimate.

But the effect is very small because the data overwhelm the judgement, as you would expect. The spread of either base distribution is much bigger than that of the YouGov result. The larger the sample the more pronounced is this effect. This means that with non-trivial survey data different initial judgements will be revised to move closer. While pro-independence and anti-independence camps may have different views about all sorts of issues which might influence the vote, if and when it comes, they should pretty much agree on what conclusions to draw from this survey.

This is especially important when contentious issues are at stake, as they are here. The message "Your judgement is swamped by the data: trust me, I'm a statistician", might not be well received. The Bayesian approach allows different opinions to be included in the analysis and their effects seen. The transparency of this process is of value (especially in these anti-elite anti-expert times).

Ecology is another area in which strong views are held. More in Chap. 14.

$$\sim \bullet \sim$$

And let us not forget grue. If you believe that the idea of grue is preposterous as a possible colour of emeralds then your base rate will be zero. You are certain grue is not the correct alternative and so when you see a green emerald you believe, as you always did, that emeralds are green (Fig. 12.9).

Fig. 12.9 Grue is implausible

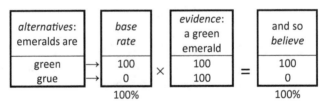

$$\sim \bullet \sim$$

Before moving on to look at more applications, a word about complex models and their solution.

In the cases we have looked at so far the base rate and likelihood distributions have been simple enough so that you could see how the calculations were done, though you had to take my word for finding probabilities from the Beta distribution. Sometimes the mathematical form of the likelihood function is such that there exists no analytical solution; there is no formula, simple or otherwise, for the answer. For example, in Chap. 14 we will look at a model with ten parameters, not the one or two we've seen so far. This is complex enough that a conveniently simple analytical approach is not possible. In these situations Monte Carlo simulation often helps.

As the name implies, the basic mechanism is as simple as spinning a roulette wheel. A standard wheel has pockets numbered from 0 to 36. Gambling aside, each spin of the wheel can be seen as a way of generating one of these 37 values. Each value is equally likely to be chosen; the wheel isn't fixed. If I wanted to pick, at random and so without bias, a day in March I could just spin the wheel. If the ball landed in a pocket labelled 0 or 32 or above I would disregard that spin and try again until I got a number between 1 and 31. That would be my randomly chosen day.

There are other ways of choosing random numbers: flipping a coin, rolling a die, choosing a card and so on. This isn't how lottery winners are picked, you may be relieved to know. In the UK the first machine for this purpose was called ERNIE, the Electronic Random Number Indicator Equipment (the acronym came first, don't you think?). ERNIE was invented by one of the Bletchley Park code breakers in 1956. The necessary randomness was provided by the behaviour of electrons in neon tubes.

Your spreadsheet almost certainly has a random number generator. This is a piece of software, not a physical device. Numbers generated by computers are only pseudo-random. They are generated by computer code and so cannot be genuinely random, but for all practical purposes they behave enough like true random numbers to be good enough. An updated ERNIE still uses physical randomness.

It may seem odd that numbers chosen at random can be used for calculating the probabilities we need, but it works. Here's how.

Figure 12.10 shows a curve, a Beta distribution just like those used with the YouGov poll. In this case the probabilities (areas under the curve) can be found fairly easily, but suppose they could not and that we wanted to know the probability that less than 20% of respondents said that they would vote Yes. Monte Carlo helps. What we need is the proportion of the area under the curve to the left of the vertical line.

Fig. 12.10 Sampling points for a Monte Carlo estimate

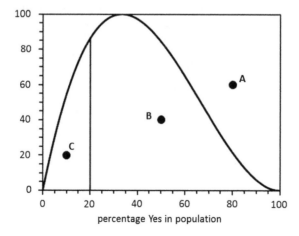

percentage Yes in population

First pick a random number from 0 to 100. You pick 80. This is the value on the horizontal axis. Pick another number, you get 60. This is the number on the vertical axis. These two random numbers define the point labelled A. This point is not under the curve and so of no use, just as a value of 34 from a spin of a roulette wheel would be no use in picking a day in March. Discard it and try again.

This time we get the numbers 50 and 40, point B. This is under the curve so keep that result.

Try again and get 10 and 20. This is point C, under the curve and also in our target area to the left of the line.

Keep going. Using a computer means you can generate lots of points. When you have all the results the ratio of the number of points under the curve and to the left of the line to the total number of points under the curve is the answer you want—the probability that no more than twenty percent of voters say they will vote Yes.

This curve is smooth but clearly this Monte Carlo method would have worked just as well with a jagged curve or any other arbitrary shape. This is why it is so useful and much used in the solution of more complex Bayesian applications.

~ • ~

So far the alternative causes or explanations we have considered have been simple single-valued propositions

> does the car belong to Stig or Jan?
> will GDP increase or decrease?
> which door to open?
> what proportion say they will vote Yes?
> does the patient have a DVT?

but sometimes the alternatives are a little more complex. For example, Antonio Pievatolo and Fabrizio Rugeri were concerned about the reliability of doors on underground trains [2]. Doors can fail in a number of ways—electrical, mechanical, pneumatic and so on. Analysis will be correspondingly detailed.

Reliability studies use time between failures as an important measure of performance. To find the average time between failures based on the evidence of a single failure time of 326 days use the Bayes Grid (Fig. 12.11).

Fig. 12.11 Bayes Grid shows the structure of analysis

alternatives: <u>average</u> time between failures	*evidence:* <u>observed</u> time between two failures ← 325 326 327 →
↑ 451.3 451.4 ↓	← likelihood ? → ← likelihood ? →

The likelihood distributions will give the probability of a failure time of 326 days if the average is, for example, 451.3 days. Considering the evidence of more failures it seems reasonable to assume independence; just multiply the likelihoods.

However, Pievatolo and Rugeri decided that the durability of train doors was a function not just of time but also of distance travelled by the train. With two variables to be estimated the model looks a little different. With only one variable, average time, the Bayes Grid analysis was, diagrammatically, two dimensional (Fig. 12.12) but now, with two variables, the Monte Carlo sampling space is three dimensional (Fig. 12.13).

Fig. 12.12 Diagrammatic representation of analysis using one variable

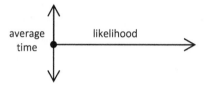

Fig. 12.13 Diagrammatic representation of analysis using two variables

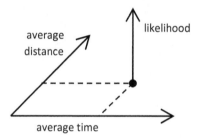

Instead of a simple curve describing our estimate of one parameter—average time or proportion of respondents—we have a three dimensional surface (Fig. 12.14).

Fig. 12.14 Surface for two variables

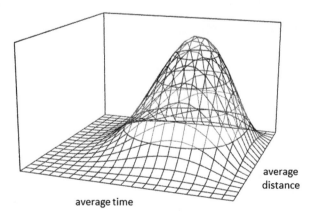

This two-variable probability distribution may have a convenient form which makes it easy to find probabilities but, if not, Monte Carlo simulation comes to the rescue again. This time, instead of picking two random numbers to find points in a square that contains a curve we need three random numbers to define a point in a cube that contains a surface. The process is otherwise the same.

This means we can answer two sorts of question. First, the volume under the surface between any given limits will give answers to questions such as "what is the probability that the average time is less than 500 days and the average distance travelled between 1000 and 1500 km?" This is the three dimensional version of the earlier two dimensional example.

Second, each randomly chosen point can be projected on to the left hand wall and on to the back wall of the cube. These two simple two dimensional curves, for average distance and average time respectively, mean that we can answer questions about each of the two variables on their own—"what is the probability that the average time is less than 400 days", for instance. This is important. The three (or more) dimensional surface shows the probability of the co-occurrence of different values of the variables but we frequently want to make probability statements about the individual variables, most often by giving an interval estimate of their value, similar to the estimate of the proportion of Yes replies in Chap. 6. This is a common use of Monte Carlo methods in Bayesian analysis, as we shall see in Chap. 12.

If you decide to read around you'll see that the particular method often used is called the Markov Chain Monte Carlo (MCMC) method.[2] There are increasingly a number of books which describe the necessary computations [4] and software is freely available, notably the winBugs software from The Medical Research Council Biostatistics Unit at Cambridge University.[3] The descriptions are, by necessity, quite technical. This outline will be enough for Chaps. 14 and 15 in the next section: Applications.

~~~ ••• ~~~

References

1. Clark SE, Godfrey RD (2009) Eyewitness evidence and innocence risk. Psychon Bull Rev 16 (1):22–42
2. Pievatolo A, Ruggeri F (2010) Bayesian modelling of train door reliability. In: O'Hagan A, West M (eds) The Oxford handbook of applied Bayesian analysis. Oxford University Press, Oxford, pp 271–294
3. Brooks SP (1998) Markov chain Monte Carlo method and its application. The Statistician 47 (1):69–100
4. Gelman A, Carlin JB, Stern HS, Dunson DB, Vehtari A, Rubin DB (2013) Bayesian data analysis. CRC Press, Boca Raton

---

[2]There are many descriptions. See, for example, Brooks [3].

[3]https://www.mrc-bsu.cam.ac.uk/software/bugs/. Accessed 23 September 2017.

Part III
Application

Chapter 13
Who Wrote That?

It is Monday 13 April in the year 1778. London. The wit, lexicographer and essayist Dr. Samuel Johnson is at dinner at Mr. Langton's house. With him, as he always is, is his friend and biographer James Boswell. Also there are Dr. Porteus, Bishop of Chester, and Dr. Stinton, a chaplain. The talk turns to the question of style, first in painting but then the styles of different authors and whether or not every writer had a distinctive style. The Bishop thought not but Johnson disagreed saying "Why, Sir, I think every man whatever has a peculiar style, which may be discovered by nice examination and comparison with others: but a man must write a great deal to make his style obviously discernible."[1]

And he was right.

~ • ~

The correct attribution of written works has long been of interest. Aristotle, in the *Poetics*, discusses which works published under Homer's name might really have been written by him (Aristotle thought the *Iliad* and *Odyssey*) [1]. More systematic study only began to be feasible with the availability of texts in collections, such as at the great library at Alexandria which was established in the third century BC. Over the centuries analyses of style began to be supplemented by some numerical measures, the idea that some authors might favour particular words and that the differential frequency of word use could help attribution.

Two types of evidence are used in attribution studies: *external* evidence such as biographical details of the possible author(s), diaries and correspondence, and *internal* evidence which is particular to a text and includes style and word or phrase usage. In using internal evidence the general approach is to describe, quantitatively or qualitatively or both, characteristics of texts of known authorship and then to

---

[1]Boswell's Life of Johnson. Chapter LXXII. Published in 1791 as The Life of Dr. Samuel Johnson LLD but available in many editions since then and commonly called just Boswell's Johnson.

© Springer International Publishing AG, part of Springer Nature 2018
A. Jessop, *Let the Evidence Speak*, https://doi.org/10.1007/978-3-319-71392-2_13

judge how closely they match the same characteristics of a text of unknown or disputed authorship. While this sounds straightforward, in principle at least, there may be difficulties. Some early writers may not have written much, or their works may have been edited or written in collaboration with others. The use of external evidence and judgements of style are likely to be much used in these cases.

Stylometric methods based on measurement and tabulation can help, of course, but need both a sufficiently large number of texts in order to decide what the style is (as Dr. Johnson well knew) and the computational ability to process these data. The availability of computers from the 1960s on meant that quantitative analysis could be used more easily and methods developed to exploit this capacity. These analyses inevitably require more data (texts, in this case). Fortunately the storage and searching of large databases of texts have greatly helped in this. Electronic downloads of books and notes are now commonplace.

As we have seen, the 1960s also saw the development of Bayesian methods in statistics. Some applications of Bayes' Rule were fairly easily implemented by hand or with simple calculators but the increasingly ubiquitous computers enabled a much wider range of problems to be addressed. It is from this period that our case study is taken.

~ • ~

The War of Independence between the American colonists and Britain ended in 1776 with victory for the Americans. But now they were no longer colonies but an independent republic just what form of republican government did they want? The America then was very different from the United States we know now being just the thirteen states on the eastern seaboard from Georgia in the south to New Hampshire in the north. Each state had its own legislature and constitution. Following victory, a Continental Congress was formed to which each state sent delegates so that a common future could be decided. The Articles of Confederation were drafted by 1777 but were not ratified by all thirteen states until 1781, which gives some indication of the tensions between the various interests. There was seen to be a need to strengthen Congress, which lacked executive authority. This question was addressed by the Federal Convention which in 1787 sent to Congress a Constitution with which to replace the Articles with the object of strengthening the role of the central government. Delaware was the first state to ratify the Constitution in December 1787 by a unanimous vote in favour, thirty votes to none. Other states followed, the last being Rhode Island in May 1790. Taking all states' votes together the Constitution was ratified by 1157 votes to 761 [2].

This period saw a vigorous and often thoughtful debate about what role central government ought to play and what should be left to the states: the Federalist and anti-Federalist viewpoints.[2] The best known example is a series of articles in the Federalist cause published to influence the ratification vote in New York (which ratified the Constitution in July 1788 by 30 votes to 27). The articles first appeared in the New York newspapers two to four times each week and were later published as a single volume. They are known as *The Federalist Papers*. They are famous for

---

[2]This issue of states' rights is still alive in the US and is common to all federal or would-be federal systems, as the European Union continues to demonstrate.

the quality and elegance of their arguments and also because of the debate over who wrote which of them.

It was a convention of eighteenth century publication that pamphlets and articles proposing often contentious arguments were published anonymously so that attention was given to the arguments themselves and not the reputation, for good or ill, of the authors (a tradition which has largely died out though some magazines maintain anonymity, *The Economist* is one). *The Federalist Papers* were published using the pseudonym Publius,[3] a name previously used by Alexander Hamilton, one of the authors. The other two authors were James Madison and John Jay. The majority of the papers were written by only one of the three rather than being a joint effort by all three. Jay wrote just five of the eighty-five papers before becoming too ill to continue. Who wrote the others, Hamilton or Madison?

~ • ~

Hamilton and Madison both favoured a strong central government and cooperated on the papers. Both went on to hold high office in the republic, Hamilton as the first Secretary of the Treasury and Madison as fourth President. But their backgrounds were quite different.

Fig. 13.1 The two authors

Alexander Hamilton 1755–1804[4]   James Madison 1751–1836[5]

---

[3]Publius Valerius Publicola was one of the earliest consuls of the Republic which followed the rule of the Kings of Rome.

[4]Portrait by Daniel Huntington. Image, US Treasury. Wikimedia Commons https://commons.wikimedia.org/wiki/File:Hamilton_small.jpg Accessed 23 S September 2017.

[5]Portrait by John Vanderlyn. Image, The White House Historical Association. Wikimedia commons https://commons.wikimedia.org/wiki/File:James_Madison.jpg Accessed 23 S September 2017.

Alexander Hamilton was born illegitimate in St Croix in the Danish West Indies, now US Virgin Islands since their sale in 1917. He started his working life as a clerk in his mother's store before moving to New York in his teens. He was Chief of Staff to George Washington in the War of Independence. When Washington became president he appointed Hamilton as the first Secretary of the Treasury of the new republic.

In 1804, some time after leaving office, Hamilton spoke against Aaron Burr (Jefferson's vice president) when Burr unsuccessfully ran for governor of New York State. They were old antagonists and Burr took exception to some remarks about himself which had appeared in the press and were attributed to Hamilton. Burr demanded satisfaction by duel which duly took place at Weehawken, New Jersey. Hamilton was killed. It was the same place where his eldest son Philip had died, also in a duel, three years earlier. Father and son used the same borrowed pistols

James Madison had a grander start in life. Born in Port Conway, Virginia he, in due course, inherited the family tobacco plantation and so a great many slaves, as was usual. He was much involved in revolutionary politics in his native state and was one of its delegates to the Continental Congress and later the Constitutional Convention. Because of his prominent role in drafting it he has become known as the Father of the Constitution.

He was an astute politician, well known for his ability to strike a necessary compromise. For example, to get the slave owning states of the South to support the Constitution he brokered a deal whereby for the purposes of representation and taxation slaves would be treated as members of the population of the state and not as the property of their owners. Five slaves were counted as three people.

He became Jefferson's Secretary of State and, on Jefferson's death, succeeded him to be the fourth President. At the end of his active political life he retired back home to Virginia.

~ • ~

Why the uncertainty about who wrote what?[6] Well, politicians don't like to be shown to be inconsistent (hard in our internet age) and so both Hamilton and Madison might have found it convenient that those papers be unattributed which expressed views they later rejected. And yet the desire to claim authorship was strong. Hamilton did leave a list but it was not drawn up calmly or openly. Two days before the fatal duel with Burr he stopped by the law office of Egbert Benson, a friend, and ostentatiously slipped his list into Benson's bookcase. The list, retrieved after the duel, gave Hamilton as the author of sixty-three papers.

After he left the presidency Madison, who had been reticent about authorship, said that he thought Hamilton may well have made mistakes in his list "owing doubtless to the hurry in which the memorandum was made out". Madison now claimed authorship of twenty-nine papers rather that the fourteen for which Hamilton had given him credit.

---

[6]A good account, from which this description is taken, is given by Adair [3].

Over the years scholarly attribution of the *Federalist* papers tended to change as the posthumous reputations of Hamilton and Madison changed. By the beginning of the 1960s the consensus was that all but twelve of the papers had been attributed

Jay	5
Hamilton	51
Madison	14
Hamilton and Madison	3
Hamilton or Madison?	12

Perhaps a little old school stylistic analysis would help. As might be expected these two men differed in how they expressed themselves, not least in how they wrote. Richard Brookhiser, in his short biography of Hamilton, [4] put the difference like this: "Jay and Madison can be epigrammatic ... or fussily formal ... Hamilton is flowing, sometimes overflowing, and agitated." This might help in deciding the authorship of the *Federalist* papers. A little quantitative analysis might help too.

~ • ~

In the early 1960s this problem of attribution attracted the attention of two statisticians, Frederick Mosteller and David Wallace (M&W for short), and it is their study[7] which is used here.

The task was to find some numerical measure to discriminate the writings of Hamilton from those of Madison in the twelve papers the authorship of which was disputed. Early thoughts that sentence length might provide that measure were quickly abandoned; both average and standard deviation (a measure of variability) were almost the same (Table 13.1).

	sentence length (words)	
	average	standard deviation
Hamilton	34.55	19.2
Madison	34.59	20.3

Table 13.1 Sentence lengths for Hamilton and Madison[8]

---

[7]Mosteller and Wallace [5]. This is a second and expanded edition of the original 1964 study *Inference and Disputed Authorship: The Federalist*.

[8]Mosteller and Wallace [5] p. 7.

A measure of style might possibly have been made based on phrases typically used by the authors but the definition and measurement problems of dealing with phrases were likely to be difficult. Using single words would be easier and was common in attribution studies. The historian Douglass Adair suggested that the differential frequency of some words might be strong indicators of authorship of the *Federalist* papers. In particular he noted that in the situations where Hamilton used *while* Madison used *whilst*.

Word use is evidence of authorship and so a Bayes analysis should help. The rate of occurrence of a word in the papers of known authorship together with an appropriate likelihood function will give the degree to which we should believe that Hamilton rather than Madison is the author of a disputed paper. (We've seen this idea before, in Table 7.4.) M&W give an example based on the usage of the word *also*. This is not a frequently used word, making up 0.031% of the text written by Hamilton and 0.067% of Madison's text: about 3 in 10,000 for Hamilton and twice that for Madison.

In analysing the *margin of error* in Chap. 6 we wanted to estimate the percentage of voters saying they would vote Yes given an assumed proportion of the total population of Scotland who would say the same. The Binomial distribution was the likelihood function. The analogous situation here is that given a known (or assumed) rate of usage of *also* the Binomial model will give the probability of finding the word used 0 or 1 or 2 or more times in a paper of a given length (the sample size). For a 2000 word paper the results are shown in Table 13.2.

alternatives:			*evidence:* occurrences of *also* in 2000 words					
author	→	rate %	0	1	2	3	...	
Hamilton	→	0.031	53.8	33.4	10.3	2.1	...	100%
Madison	→	0.067	26.2	35.1	23.5	10.5	...	100%
probability of Hamilton			67.3	48.7	30.5	16.9	...	

Table 13.2 Likelihoods for the occurrence of *also*[9]

---

[9]Mosteller and Wallace ([5], Table 3.1-1, p. 53).

For example, if *also* doesn't appear at all there is a 67.3% probability that the author was Hamilton, odds of about 2:1. If *also* appears three times there is a probability of just 16.9% that Hamilton was the author, odds of about 5:1 in favour of Madison. This shows that Bayes' Rule certainly gives a way of making attributions, which is what M&W wanted, but before it could be used for the *Federalist* problem three questions had to be answered and these are shown in a Bayes Grid (Fig. 13.2).

Fig. 13.2 Bayes Grid structures analysis for *also*

alternatives:		evidence: occurrence of *also*					base rate
author → rate %		0	1	2	3	...	
Hamilton → "about 0.031"		← likelihood ? →					?
Madison → "about 0.067"		← likelihood ? →					?

The three questions were

what base rates should be used?
is the Binomial model adequate for these likelihoods?
how accurately can the word rates be estimated?

and also there was the important question

which words should be used as evidence?

Finding good answers to these questions is what most of M&W's study describes. They were working in the 1960s when this approach to attribution was new and so had to establish most of the details of the application. Seeing how they answered the questions raised by the Bayes Grid will show decisions typically necessary for the basic idea to be used for realistically complex problems.

~ • ~

What Base Rates Should Be Used?
M&W were statisticians not historians, linguists or experts in the study of literary style. They felt uneasy about giving base rates. These have to be based on judgement or data which are independent of the evidence used for the rest of the analysis. M&W did not feel competent to make such judgements even though these are needed for the application of Bayes' Rule. So they didn't.

There are a number of responses to the problem of being so unsure about giving values for base rates, or any other parameters, that you feel uncomfortable giving any values at all. For probabilities, giving flat distributions, all probabilities equal, is one way of encoding maximum ignorance (Chap. 10). Alternatively, many people find sensitivity analysis useful. Changing the values of parameters one at a time and seeing the effect can help. It may be that the results are sufficiently robust as to be useful. Alternatively, if not, the sources of imprecision are identified in the hope that more thought or data will improve the situation. If this is not possible the analysis can support only an appropriately tentative conclusion. M&W did not see themselves as doing any of these.

They postponed consideration of base rates until after they had made attributions of authorship using an analysis based on likelihoods alone. Although postponing the, for them, uncomfortable task of giving base rates looks as if no base rates are given, we have seen that what this really means is giving equal rates, 50:50 in this case—a numerical expression of maximum ignorance, which is just how the two statisticians felt.

M&W, or you or I, could then consider how strong our views, our base rates, would have to be to upset the result in favour of the other author. In the illustration, if the word *also* was found three times in a 2000 word document the odds, based on that evidence alone, are 5:1 in favour of Madison, a probability of 83% that Madison was the author. If your initial feeling was that Madison was the author you have support for your view. But if you felt that Hamilton was the author you now have evidence to the contrary. The strength of the evidence is described numerically but your initial feeling is not, which is why you did not want to give base rates. (We used this argument in Fig. 10.1 to think about Steve's job.)

While it can be difficult to give probability estimates of what you believe it is generally much easier to react to a proposed value, especially if that value is very large or very small. In the extreme, the evidence may point so overwhelmingly to one conclusion that your initial judgements would have to be similarly extreme, but in favour of a different alternative, for a decision in favour of that different alternative to be justified.

What is the critical value when deciding authorship? Having seen *also* three times we would need to believe that, based on judgement unrelated to the analysis of the text, the probability that Hamilton was the author was more than 83% to justify deciding in his favour (Fig. 13.3).

Fig. 13.3 Base rates needed for equal beliefs in authorship

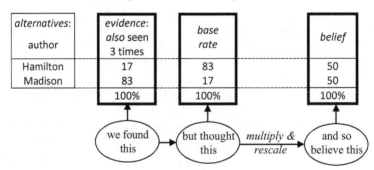

Using this approach cautious statisticians postpone making that awkward prior judgement about authorship and stick to that quantitative analysis in which they are expert. Consumers of that analysis, whether you or me or anybody else, are then faced with a recommendation that

> if you judge that the probability that Hamilton wrote this text is greater than eighty-three percent you are justified in believing him to be the author, otherwise you should believe that Madison wrote it

or, equivalently,

> if you judge that the odds that Hamilton wrote this text are better than five to one you are justified in believing him to be the author, otherwise you should believe that Madison wrote it

either of which should be easier than asking you for base rate probabilities.

We have been used to thinking about Bayes' Rule like this

$$\textbf{belief } \text{is proportional to } \textbf{base rate} \times \textbf{likelihood}$$

so that, because we read English from left to right, the implication is that we think first about our base rates because they describe the context of the problem, and then look at the evidence. But here we reverse that order and think like this

$$\textbf{belief } \text{is proportional to } \textbf{likelihood} \times \textbf{base rate}$$

to ease the difficulty, if there is any difficulty, of giving judgemental base rates.

When the base rates are found from data, as with medical diagnosis, the difference in these two approaches is immaterial but when the interpretation of results relies on the judgement of the user of the analysis it may be more useful to do the calculation first using maximum ignorance base rates. Even if these calculations are quite complex, as with M&W's work, the results are easily stated as probabilities or odds about which non-statisticians may have a view.

And so M&W used this wait-and-see strategy both in the hope (I think) of possibly not having to make a difficult judgement and also to leave scope for interpretation by others.

~ • ~

Is the Binomial Model Adequate for These Likelihoods?
In the illustration likelihoods were given by the Binomial distribution. This is a model commonly used for this sort of problem because it relies on just a few assumptions which are frequently met, in this case that the probability that any word in a text is *also* is the same in all places in the text and that occurrences are independent. It is common that the number of times the word is seen in each of a number of texts is well enough predicted by the model for the predictions to be useful.

The Poisson distribution is a simpler version of the Binomial.[10] The basic assumptions are the same but only the average number of occurrences, not the sample size and rate, is needed. This is often more convenient and was used by M&W.

It always makes sense to check if the model does, in fact, provide sufficiently accurate forecasts. M&W tested the adequacy of the Poisson distribution by getting a number of blocks of text of no less than a hundred words, and mostly more than a thousand, from papers and other material known to have been written by Hamilton or by Madison. The number of times a given word was seen in each of the blocks was found. This was the likelihood distribution to be modelled. For example, the black bars in Fig. 13.4 shows the distribution for the word *an* in two hundred and sixty-two bocks known to have been written by Madison. In seventy-seven blocks *an* appeared only once.

Fig. 13.4 Comparison of actual and estimated distributions of *an*[11]

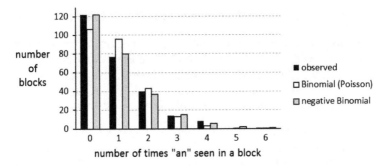

The white bars to the right show the distribution predicted by the Poisson model. Ninety-six blocks contained just one occurrence of *an*. This was the general picture; the model tended to underpredict the number of papers in which the word occurred not at all and also where it occurred a lot and consequently to overpredict the numbers of papers with just one or a few occurrences. In short, the observed distribution showed more variability than the Poisson predictions.

M&W reasoned that there must be some other source of variation than that described by the model. The Binomial/Poisson model makes the critical assumption that for any block of words, or subsets of that block, the rate for any word is the same. It doesn't matter if we take the first hundred words or the last or any other, the

---

[10]Specifically, when the sample size is large and the rate small. Both conditions hold for these documents.

[11]Mosteller and Wallace [5] Table 2.3-4, pp. 32–33.

average rate with which a word occurs is the same. But can that be true? It is surely plausible that words may be more common in some sections than others depending on the subject of the writing. An even bigger effect for M&W arises because they took blocks not just from different sections of a paper but from different papers too. The Negative Binomial distribution, closely related to the Binomial model, better allows for this variability. The predictions of this model are much closer to the observed values, as Fig. 13.4 shows, and so was used by M&W.

~ • ~

How Accurately Can the Word Rates Be Estimated?

Being careful statisticians M&W knew that finding word rates for the Poisson model, or values for the parameters of the Negative Binomial model, had inevitably to be based on samples of the writing of Hamilton and Madison (the undisputed *Federalist* papers, most obviously) not the whole body of their work. Rather than knowing that the rate for Hamilton was exactly 0.031 occurrences of *also* per thousand words all that could be said was that it was "about 0.031". Since the value is estimated from sample data "about 0.031" means finding a probability distribution to measure the margin of error of the estimate. Once that distribution is found the likelihood for each possible rate, and the probability of that rate, can be used to give a probability distribution of odds or belief probabilities that Hamilton was the author. This distribution will be a Beta distribution, just like the one used in Chap. 6.

In just the same way there will be a margin of error for Madison's rate of "about 0.067". Combining the two margin of error distributions gives probabilities of all pairs of rates, one for Hamilton and one for Madison, and the corresponding likelihoods and belief probabilities.

The same approach is used to find the margin of error for the parameters of the Negative Binomial model.

Putting all this together is not simple. The result for each paper is not a single figure for probability but a distribution describing the margin of error of that estimate, just as in Chap. 6 we had the probability that the proportion of voters was between two limits. This means that users of the report will be looking at probabilities of probabilities, which is not for everyone. Alternatively, report the average. This is what M&W did.

The mathematics of all this is not for this book but you can see how Bayes thinking about likelihoods was the foundation of the analysis.

~ • ~

Which Words Should Be Used as Evidence?
Early studies identified *marker words*, those which discriminate between Hamilton and Madison. Table 13.3 shows the rates of occurrence per thousand words of four markers for the *Federalist* papers.

author	marker words				no. of words
	enough	*while*	*whilst*	*upon*	
Hamilton	0.59	0.26	0	2.93	45,700
Madison	0	0	0.47	0.16	51,000
Disputed	0	0	0.34	0.08	23,900
Joint	0.18	0	0.36	0.36	5,500

Table 13.3 Rate of occurrence per thousand words of some marker words[12]

M&W comment that "These data show rather clearly that the disputed papers as a whole are Madisonian, but in this form they cannot settle the papers singly … What is wanted from these and other data is, for each paper, a good measure of the weight of evidence toward Madison or Hamilton."[13] They wanted a better measure because, although the number of words available for the analysis is in the thousands, the rate of occurrence of the marker words is very small. The *Federalist* papers are each about 2000 words long so whether the appearance or non-appearance of one of these marker words is due to who wrote the paper or just the small sample provided by the text is not easy to decide. *War and Peace* they are not.

The lengths of the twelve papers were fixed. To increase discrimination M&W needed more evidence and this meant using more than just one word so that "the evidence is overwhelming, though no one clue is".[14] While this increased the complexity of their model it is in principle no different from other applications of Bayes' Rule: more evidence leads to better discrimination. But which words to use?

To choose the most useful needed an analysis of their occurrence in papers known to have been written either by Hamilton or by Madison. But the *Federalist* papers of known authorship are not numerous when considered as a statistical sample. This is so particularly for Madison. M&W decided to augment the sample by using other papers written at about the same time as the *Federalist* set and on similar topics. These they called exterior papers. The result was forty-eight papers by Hamilton and fifty by Madison.

---

[12]Mosteller and Wallace [5] Table 1.4-2, p. 11.

[13]Mosteller and Wallace [5] p. 12.

[14]Mosteller and Wallace [5] p. 10.

In thinking about which words to use M&W distinguish function words and contextual or content words. Function words (such as prepositions and pronouns) are those which are expected to occur whatever the topic being discussed and so might be thought characteristic of the prose style of the author. For example, the use of *by* is shown in the left half of Table 13.4.[15] The column for each author shows the number of papers by density of occurrence of the word.

function word *by*			contextual word *war*		
rate per thousand	author Hamilton	author Madison	rate per thousand	author Hamilton	author Madison
1 - 3	2		0 exactly	23	15
3 - 5	7		0 - 2	16	13
5 - 7	12	5	2 - 4	4	5
7 - 9	18	7	4 - 6	2	4
9 - 11	4	8	6 - 8	1	3
11 - 13	5	16	8 - 10	1	3
13 - 15		6	10 - 12		3
15 - 17		5	12 - 14		2
17 - 19		3	14 - 16	1	2
	48	50		48	50

Table 13.4 Function words discriminate more than contextual words

Both authors used *by* quite frequently, though Madison more than Hamilton and so *by* discriminates quite well: lower frequency of use indicating Hamilton as the more likely author and higher frequencies indicating Madison.

Contextual words (nouns, verbs, adjectives and adverbs), on the other hand, are less common for both authors and highly dependent on the subject of the paper. Many papers may not contain the word at all. The pattern of use of *war* in the right half of the table clearly shows this.[16] The effect of the topic being discussed swamps the stylistic preferences of the authors. Whoever is writing, if the paper is about war then *war* will appear as needed by the discussion. If war is not the subject of the paper *war* may well not appear at all. Contextual words are less likely to provide evidence of authorship.

---

[15]Mosteller and Wallace [5] Table 2.1-1, p. 17.
[16]Mosteller and Wallace [5] Table 2.1-3, p. 19.

This gives just a flavour of the issues which needed to be resolved. After much careful analysis M&W chose these thirty words:

according	*both*	*enough*	*on*	*to*
also	*by*	*innovation(s)*	*particularly*	*upon*
although	*commonly*	*kind*	*probability*	*vigor(ous)*
always	*consequently*	*language*	*there*	*while*
an	*considerable(ly)*	*matter(s)*	*this*	*whilst*
apt	*direction*	*of*	*though*	*work(s)*

Having chosen these words as their evidence and using the Bayesian model M&W were able to decide the authorship of the disputed *Federalist* papers. Using thirty words rather than just one complicates the analysis but provides a much better evidence base for the final decision.

~ • ~

The results were overwhelmingly in favour of Madison. Table 13.5 shows, for each of the twelve disputed papers, the probability that they were written by Madison. The papers are shown in order of increasing probability. M&W gave maximum and minimum estimates.

Federalist Paper number	probability (%) range that Madison was the author	
	low estimate	high estimate
55	97.069	99.632
56	99.451	99.978
49	99.963	100.000
50	99.982	100.000
54	99.982	100.000
58	99.989	100.000
53	99.993	100.000
52	99.993	100.000
62	99.995	100.000
57	99.995	100.000
63	99.998	100.000
51	100.000	100.000

Table 13.5 Very strong support for Madison[17]

---

[17]Re-expressed from Mosteller and Wallace [5] Table 3.7-2, p. 87.

For papers 49 onward the minimum is 99.96%. Even for paper 55 the low estimate is over 97%. You would have to have as base rate a probability in favour of Hamilton greater than this value to sustain a belief that that the author was not Madison. It seems highly unlikely, to put it mildly, that you would have grounds for such a strong belief. This is just what M&W thought might be the case when they decided not to specify base rates but to wait and see what their analysis showed: reaction rather than specification.

~ • ~

The main interest of M&W was methodological. Their study was thorough and can be summarised in a number of ways. Here, I have emphasised how the basic structure of their analyses relies on those simple elements of Bayes' Rule that we have met before. The extensions needed by M&W have been sketched to show some of the model building necessary, but the basics are unaltered.

Previous attributions, notably by Alvar Ellegård [6], had used statistical methods but were not Bayesian. M&W's analysis is generally taken to have settled the matter to the extent that correctly attributing the disputed papers to Madison is taken as a benchmark for testing other methods [7–10], though the possibility that some of the disputed papers were, to some extent, written jointly by Hamilton and Madison has been raised [11].

~ • ~

The need to decide who wrote what is widespread and growing. Forgery and plagiarism are pretty obvious areas where some detective work, aided by a little statistics, is needed. Perhaps not so obvious is trying to find how much of the text of bills presented by US legislators is actually provided by lobbyists. The University of Chicago's Legislative Influence Detector[18] can help [12].

Not all the methods are Bayesian, but the work of Mosteller and Wallace prepared the ground.

~~~ ••• ~~~

References

1. Love H (2002) Attributing authorship: an introduction. Cambridge University Press, Cambridge, p 15
2. Dry M (1991) The debate over ratification of the constitution, Ch. 48. In: Greene JP, Pole JR (eds) The Blackwell encyclopaedia of the American revolution. Blackwell, Cambridge Mass, p 471–486.

---

[18]Sunlight Foundation. Tracing Policy Ideas From Lobbyists Through State Legislatures. https://dssg.uchicago.edu/project/tracing-policy-ideas-from-lobbyists-through-state-legislatures/?portfolioID=15901. Accessed 23 September 2017.

3. Adair D (1944) The authorship of the disputed Federalist papers. William Mary Q 1(2): 97–122 (1(3): 235–264)
4. Brookhiser R (1999) Alexander Hamilton, American. Touchstone, New York
5. Mosteller F, Wallace DL (1984) Applied Bayesian and classical inference: the case of the Federalist papers. Springer, New York
6. Ellegård, A (1962) A statistical method for determining authorship: The Junius Letters, 1769–1772. Gothenberg Studies in English No. 13. Acta Universitatis Gothoburgensis, Gothenburg
7. Bosch RA, Smith JA (1998) Separating hyperplanes and the authorship of the disputed Federalist Papers. Am Math Mon 105(7):601–608
8. Fung G (2003) The disputed Federalist papers: SVM feature selection via concave minimisation. Proceedings of the 2003 conference of diversity in computing. ACM Press, New York, pp 42–46
9. Holmes DI, Forsyth RS (1995) The Federalist revisited: new directions in authorship attribution. Lit Linguist Comput 10(2):111–126
10. Savoy J (2013) The Federalist Papers revisited: a collaborative attribution scheme. Proc Am Soc Inf Sci Technol 50(1):1–8
11. Collins J, Kaufer D, Vlachos P, Butler B, Ishizaki S (2004) Detecting collaborations in text: Comparing the authors' rhetorical choices in The Federalist Papers. Comput Hum 38(1):15–36
12. Rutkin A (2015) Law influencers unmasked. New Sci, 14 Nov, p 22

Chapter 14
Wood, Trees and Wildlife

The pressures of a growing population have reduced the area of the planet covered by forests from 4128 million hectares in 1990 to 3999 million hectares in 2015, a reduction by an area about the size of South Africa [1]. This global decline is important for a number of reasons, not least because of its effects on biodiversity and climate. A minority of forest is used for commercial purposes. In 2015 global trade (exports) in wood and wood products was worth US$226 billion [2]. The management of forests for profit is different from management to meet purely ecological objectives but in both cases knowing the number and state of the trees is important.

Ecological models are helpful in a number of aspects of forest management, two of which—how much sellable wood a forest contains and what are the impacts on forest wildlife of commercial logging—will be used to show how.

Each illustrates a different strength of adopting a Bayesian approach; how to deal with difficult likelihoods and how to use base rates when strongly held views might affect a decision.

~ • ~

The owners of a commercial forest will want to know the value of their asset; how much wood they have in their forest and how much it is worth. The inventory which gives this information is called a timber cruise. Money value will be a function of the species of tree and the purpose for which the harvested wood might be used—for sawnwood or wood pulp, for instance. The assessment then rests on just how much wood is in the forest. This is measured as cubic metres of merchantable wood. Finding the volume of each of a sample of trees in the forest gives an estimate for the whole forest.

How to find the volume of wood in a tree? Think of the trunk as a tall tapering tube, like a very slender cone. To find the volume of a cone requires only the base

© Springer International Publishing AG, part of Springer Nature 2018   155
A. Jessop, *Let the Evidence Speak*, https://doi.org/10.1007/978-3-319-71392-2_14

diameter and height, and so knowing the height of the tree and its base diameter should enable a good estimate of its volume to be made. By convention, the diameter is measured not where the trunk emerges from the ground but at about chest level. Some form of calliper, such as that shown in Fig. 14.1, makes this easy to do. Well, almost. There is no universally agreed height above ground at which this measurement, the diameter at breast height (Dbh), should be made. But the differences are too small to have a material effect.

Fig. 14.1 Electronic tree caliper[1]

In principle, measuring height is also easy. You have probably seen a picture of a painter holding a pencil vertically at arm's length to find the relative heights of objects in a drawing or painting; a building, a hill, even a tree. In this very low-tech method hold a measuring stick (pencil, ruler) in front of you and adjust how far away you are from the tree until the bottom and top of the tree and the bottom and top of your stick are aligned. If you are twenty times further from the tree than your stick is from your eye then the tree is twenty times taller than your stick.

Alternatively use a clinometer. This measures the angle between the horizontal and your sight line to the top of the tree. Then, with a little trigonometry, it is easy to find a factor by which to multiply the distance to the tree to get its height.

Or download an app for your cell phone.

You will have to be clear about the location of the top of your tree. Only measure as far up the trunk as the tree is wide enough to be commercially useful for whatever purpose you have in mind. This is the merchantable height.

---

[1](Image, Claudius. Wikimedia Commons. https://commons.wikimedia.org/wiki/File:Electronic_caliper.jpg Accessed 24 September 2017)

Helpful folk have made estimates of the volume of wood given the type of tree you have measured. For example, Table 14.1 shows a fragment of a table provided in the 1970s by the Forest Service of the US Department of Agriculture:

Dbh	Merchantable height in feet						
(inches)	10	15	20	25	30	35	40
15	.322	.438	.552	.659	.758	.849	.932
16	.368	.501	.631	.753	.867	.971	1.065
17	.417	.568	.716	.855	.983	1.101	1.208
18	.470	.640	.806	.963	1.108	1.241	1.361
19	.526	.717	.903	1.078	1.240	1.389	1.524
20	.586	.798	1.005	1.200	1.381	1.547	1.697

Table 14.1 Gross weight in tons for hard hardwoods ([3], Table 10, p. 14)

Nowadays, of course, this information will be provided in some useful software.

All of this is just what you want, provided you can see the top of your tree. No problem for a lonely tree in the middle of a field but no good in a densely packed forest where the tops of the trees merge to form a canopy.

What to do?

It seems likely that, for any given species, there is a relation between height and diameter, for any species, taller trees have fatter trunks (Fig. 14.2).

Fig. 14.2 Linear relation between tree height and diameter

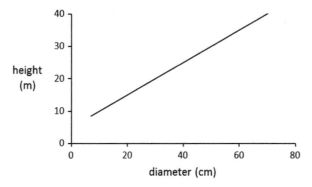

Once we have this simple model the height can be estimated from the diameter and the volume of merchantable timber found.

Later in this chapter we'll see how Bayes' Rule helps us to find an appropriate relation between height and diameter. But first, here is a simple illustration to show some of the modelling issues.

~ • ~

To find the right straight line for a given species of tree collect some data. The points on the graph in Fig. 14.3 are the diameter and height measurements for the nine trees in a (small) sample.

Fig. 14.3 Linear model with data points

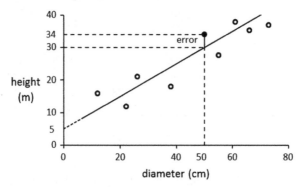

The straight line is fixed by choosing values for the slope of the line, its gradient, and the intercept. The slope of this line is 0.5; an increase in diameter of 20cm is associated with an increase in height of 10m. The intercept is the value of height at zero diameter. No such trees exist, of course (no diameter, no tree) but the intercept fixes a particular line of all those parallel lines with the same slope. Think of the intercept as a necessary parameter. In this case the intercept is 5m. To predict the height of a tree just halve its width and add five.

The data do not fall conveniently in a straight line, they never do. But there is clearly an upward relation. Using this straight line model will give pretty good estimates of height from diameter but there will inevitably be an error such as that shown. The solid dot is for a tree 34m high with a diameter of 50cm. Using the straight line model, the predicted height is 30m, an error of 4m for this tree.

A standard statistical method called regression analysis finds the values of slope and intercept which minimise the aggregate error to give a line of best fit.[2] The analysis also gives the distribution of errors to show how good an estimate we have.

---

[2]Regression analysis fits the line which minimises the sum of squared errors. Squaring prevents positive and negative errors, overestimates and underestimates, from cancelling each other out. This method is called least squares regression. You will be able to do this on your spreadsheet.

Imagine that the sample had included hundreds of trees all with diameter 50cm. It is likely that the plot of the distribution of heights would look like the one shown in Fig. 14.4.

Fig. 14.4 Normal distribution of tree heights

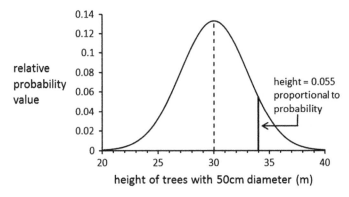

This shape occurs so often that it is called the Normal distribution.[3] Tree heights are clustered round the average, 30m. Tall trees and short trees are less common, and get rarer the taller or shorter they are.

Just like the Beta distribution used to describe the voting intentions of the people in Scotland in Chap. 6 it is the area under this curve which gives probability. The height is a value proportional to probability for any given diameter, so 34m *exactly*, plus or minus nothing. No measurements are that precise but the height of the curve gives the *relative* value of probability for any diameter. This will be useful.

One final word about this Normal distribution. You can see that just about all trees have heights within 9 m of the average (only a quarter of one percent of trees have heights outside this range). This gives a measure of variation or spread, which will be different for different types of tree. One sixth of that range, 3m in this case, is a conventional measure of spread called standard deviation.[4]

---

[3] Also known as the Gaussian distribution, named for the German mathematician Carl Friedrich Gauss. Your spreadsheet will have the necessary functions.

[4] Standard deviation is a widely used measure. For distributions other than the Normal do not just divide range by six. Consult a stats book.

This spread has two aspects. Most obviously, it describes the variation natural in many systems—trees, manufactured objects, people. But it also describes the error we might make by using the average as a prediction (Fig. 14.5).

Fig. 14.5 Normal distribution of prediction error

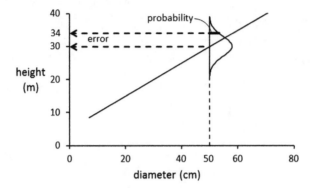

Given a reasonable number of data (a lot more than nine) regression is a good way of fitting a useful model. The simple linear model is fixed if we know three things: intercept, slope, and error spread. The regression model gives all three.

~ • ~

But you may not have any data, or not so many that you are confident that fitting a regression model would give reliable results that you would be happy to use. You may be familiar with other models like this for the prediction of tree height and so, from your experience, are confident that the intercept is either 5 or 7 and the slope is either 0.5 or 0.6. (This is an artificial illustration, remember). There are four possible straight line models (Table 14.2).

model	intercept	slope
A	5	0.5
B	7	0.6
C	5	0.6
D	7	0.5

Table 14.2 Four possible models

Which model to choose? If we have no preference we could make four predictions of height and take the average.

But we have the heights and widths of a few trees as evidence to assess the models' performance and so decide between them. For example, our 34m high tree with a trunk 50cm in diameter can be plotted on the same graph as the four models (Fig. 14.6).

Fig. 14.6 Using four models to predict

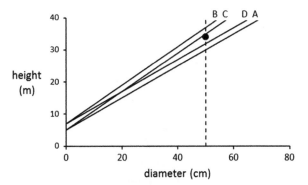

This one observation provides some evidence to help. Model C, with an intercept of 5 and slope of 0.6, gives a good prediction of the height of our tree, overestimating by just 1m. It would be foolish to reject the other three models on the evidence of just one tree. Bayes' Rule provides a framework for judging by how much we should now prefer one model to another; what degree of belief we should have that model C is the true (more true, at any rate) description of the relation between diameter and height. The Bayes Grid shows what we need to know (Fig. 14.7).

Fig. 14.7 Bayes Grid shows shape of analysis

alternatives:			base rate	evidence: tree height
model	→	prediction		... 33 34 35 ...
A	→	30	?	← likelihood ? →
B	→	37	?	← likelihood ? →
C	→	35	?	← likelihood ? →
D	→	32	?	← likelihood ? →

If we have no views on the parameter values, use equal base rates.

Likelihood distributions give probabilities of what we see as evidence: how likely is a tree 34m high given that its diameter is 50cm and model A is the true relation between height and width? We already have this. The Normal distribution of errors shows that using model A the prediction is a height of 30m and the Normal error distribution shows that the probability of a height of 34m is proportional to 0.055. To keep things simple assume the same likelihood distribution for each model, differing only in the average, the predicted value (Fig. 14.8).

Fig. 14.8 Four error distributions

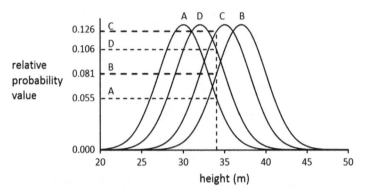

Putting all this information in the Bayes Grid (Fig. 14.9) confirms what the simple predictions show—the prediction of model C is the best and that of model A the worst—but we can now see how strongly we should believe in each of the models.

Fig. 14.9 Bayes Grid analysis for the four possible models

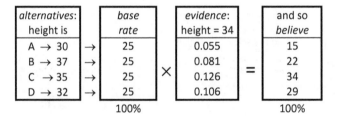

alternatives: height is		base rate		evidence: height = 34		and so believe
A → 30	→	25		0.055		15
B → 37	→	25		0.081		22
C → 35	→	25	×	0.126	=	34
D → 32	→	25		0.106		29
		100%				100%

This is just a sketch to show how some Bayes thinking helps in deciding between models. The straight line model was simple. This Bayesian approach to parameter estimation comes into its own with more complex models. We'd need more than one lonely tree as evidence, of course.

In practice, the possible values of the parameters would not be so limited. Rather than describing the intercept as being either 5 or 7 we might want to say that based on experience elsewhere the intercept is somewhere between 3 and 12 and so have a base rate probability distribution for all values in this range, for slope and standard deviation of errors too.

~ • ~

The longleaf pine (*Pinus palustris*) is found in the south eastern United States. Originally valued for its resin it is now used for lumber and pulp. These trees can be long lived and grow to about 50m. As Fig. 14.10 shows, the trunk is slim and does not taper much. The needles are longer than those of other pines, hence the name.

Fig. 14.10 Long leaf pines[5]

Once plentiful, poor management has resulted in a decline of the tree to the extent that the area in which it is dominant is now only about three percent of what it was at its peak.

---

[5](Image, M Fitzsimmons. Wikimedia Commons. https://commons.wikimedia.org/wiki/File:Trail_in_Forest.JPG. Accessed 24 September 2017)

The life of a longleaf pine can be thought of as occurring in a number of stages. The Longleaf Alliance identifies six: seed, grass, bottlebrush, sapling, mature, old growth, and then death.[6]

William Platt and his colleagues from Florida State University in Tallahassee measured the characteristics of a random sample of nearly four hundred longleaf pines in an 80ha tract, the Wade Tract, of Thomas County in Georgia [4]. The aim of their study was not to calibrate a diameter/height model but their data were used to do just that by Brian Beckage of the University of Vermont working with Platt and others [5].

In the illustration above a straight line model was used. It is the simplest form. But trees do not grow forever to infinite height and girth, there must be a limit (Fig. 14.11).

Fig. 14.11 There is a limit to the height of a tree

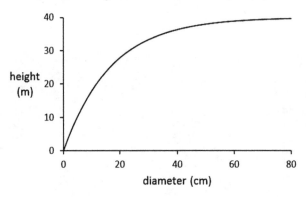

The data from the Wade Tract shows this tendency (Fig. 14.12).

Fig. 14.12 Data from the Wade Tract[7]

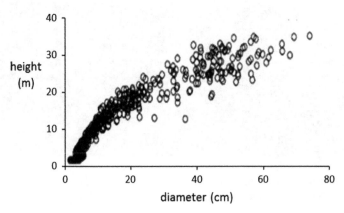

---

[6]The   Longleaf   Alliance.   Alabama,   USA.   http://www.longleafalliance.org/   Accessed 24 September 2017.

[7] (Beckage et al. Figure 4(a) p. 462. Reproduced courtesy of *The New Phytologist*, Wiley).

This is fine, but what about the six different stages of growth suggested by the Longleaf Alliance? It is at least plausible that in growing from one stage to the next the relation between diameter and height also switches from one form to another. A smooth curve does not describe these sharp changes. Beckage and his colleagues simplified the six stages to three: the grass stage, in which juvenile trees grow in diameter but not much in height; the adult or middle stage, in which both diameter and height grow steadily; the final stage, in which the growth enters the canopy, light is reduced and so is the rate of growth of height. It was assumed that at each stage the relation between diameter and height was linear. The points of transition between stages are called change-points and occur at diameters cp_1 and cp_2 in Fig. 14.13.

Fig. 14.13 Change-point model

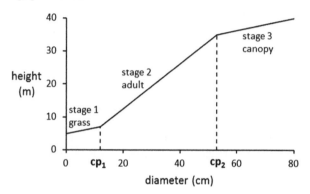

Although the three sections are just straight lines the introduction of the change-points complicates the model. The straight line models can only be fitted and the error distribution found once the change-points are known. But these points and the lines between them must be found simultaneously. Simple regression model fitting will not work, but a Bayesian approach will. The spread of errors is needed for the likelihoods and so was treated as a model parameter. The nine parameters to be estimated were

for each stage:	the slope and the spread of errors
for stage 1:	the intercept
for each change-point:	the diameter at breast height

Very young trees aren't tall enough to reach to breast height. In those cases the diameter close to the ground was measured and an allowance added to give an estimate of the equivalent diameter at breast height. This tenth parameter was also included in the Bayesian analysis by Beckage and his colleagues.

They wanted the results of the analysis to depend primarily on the data rather than on their judgements and so chose quite flat and non-informative base rate distributions. A number were tried. Examples are shown in Figs. 14.14 and 14.15.

Fig. 14.14 Base rate distribution for slope

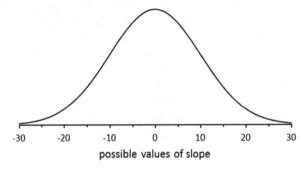

possible values of slope

Fig. 14.15 Base rate distribution for change-point

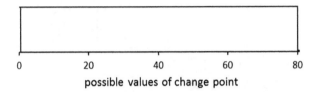

possible values of change point

The calculations were not simple and so a Monte Carlo analysis, like the one described in Chap. 12, was made using samples of 40,000. The results are shown in Table 14.3.

parameter	parameter estimate		
	low	average	high
intercept	-0.0257	-0.0169	-0.00803
slope stage 1: grass	0.0493	0.0553	0.0612
stage 2: adult	1.11	1.23	1.35
stage 3: canopy	0.337	0.399	0.444
error standard deviation			
stage 1: grass	0.021	0.023	0.026
stage 2: adult	0.891	1.07	1.26
stage 3: canopy	3.08	3.34	3.73
diameter at change-point			
cp_1: juvenile to adult	2.52	2.58	2.61
cp_2: adult to canopy	11.7	13.8	17.5

Table 14.3 Results of change-point analysis[8]

---

[8]Beckage et al. [5], Table 1, p. 462).

The distributions which these values summarise are quite precise. The ranges described by the low and high values are ninety-five percent intervals. For example, we are justified in believing with ninety-five percent probability that the slope for stage 2 lies between 1.11 and 1.35.

The graph (Fig. 14.16) using average values for intercept, slopes and change-points (shown by dashed lines) shows a good fit to the data.

Fig. 14.16   Change-point model fitted to the Wade Tract data[9]

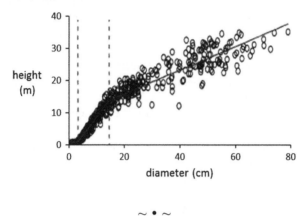

~ • ~

Beckage and his colleagues wanted a model which recognised the stages in the life of the longleaf pine. This complicated the analysis because it required the simultaneous estimation of where each section begins and ends and also the slopes in between. Using a Bayesian approach overcame this problem.

As we have seen before, in the analysis of the *Federalist* papers in Chap. 13, a cautious approach was taken to the specification of base rates.

Deciding the number of change points is also a judgement. Following Platt, Beckage and his colleagues used three stages, and so two change points, rather than six or some other number, a decision influenced by the original data collection. In other cases the number of change points have been limited so that only "larger and more abrupt changes are highlighted" [6], a more pragmatic argument.

The estimation of timber volumes is quite a technical task. There is no reason to think that there is much dispute about the basic purpose, though different models may have their enthusiasts. But some consequences of logging may be viewed quite differently, even antagonistically, by different people. There may be strongly held but different views, as the next case shows.

~ • ~

We have seen that in a wood or forest the topmost growth of trees forms a canopy and that this makes it more difficult to estimate tree height. The canopy also reduces

[9](Beckage et al. Figure 4(b). p. 462. Reproduced courtesy of *The New Phytologist*, Wiley)

the light that reaches the ground beneath it. Changes in light level lead to changes in the habitats of plants and animals and birds that live in the forest. It is easier to assess the impact of the canopy by studying what happens when some is removed rather than seeing what happens as it grows (it's quicker). Francis Crome and his colleagues of the Commonwealth Scientific and Industrial Organisation (CSIRO) in Australia did just that [7].

A rainforest on the Windsor Tableland in Queensland was divided into two areas, an unlogged area of 45ha and a logged area of 19ha managed by the Queensland Forest Service. Trees removed during logging altered the light climate in the forest and so too did the tracks needed to get to the trees to remove the logs. Logging eliminated twenty-two percent of the canopy.

Birds and small mammals living in the forest were surveyed several times before and after logging. Birds were caught in nets and marked with metal leg bands and then released. Mammals were similarly caught and tagged. There were nine survey sites in each of the logged and unlogged areas. The number of birds and mammals trapped gave a measure of their abundance.

The changes in habitat brought about by logging may either increase or decrease abundance depending on whether the changes are favourable or not for the species in the forest. The change in capture rates was used to measure the size of the logging effect. A value of 1 meant no change, a value of 0.75 meant a twenty-five percent reduction, a value of 1.25 meant a twenty-five percent increase.

For alternative values of the size of the logging effect a likelihood function was used which depended on the difference in catches before and after logging and the time that logging occurred.

One of the species recorded was Bower's Shrike-thrush (*Colluricincla boweri*). The results for this bird are used here as illustration.

The size of the logging effect could be any non-negative value, though not much greater than about three. With a flat distribution for base rates (all values equally likely) the Bayesian analysis gave the result shown in Fig. 14.17.

Fig. 14.17   Estimate of logging effect using flat base rate distribution

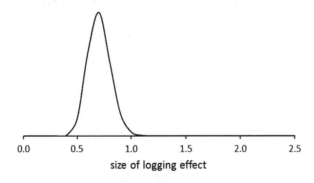

We are justified in believing that the logging effect is between 0.4 and just over 1. The most likely value is about 0.7, a reduction in abundance of thirty percent. Logging has a negative effect for this Shrike-thrush.

But what about the opinions of those with an interest in many ecological issues? There are likely to be sharply different views about the effects of logging. It is not difficult, in principle, to incorporate these views as different base rate distributions (as in Fig. 12.8). But how to get the views?

Fifteen people—loggers, conservation activists and members of the public—were asked what they thought was the size of the logging effect. They were asked seven questions

1. Choose a level of impact (percentage increase or decrease) so that there is a 50% chance that the effect will be below this level (and therefore a 50% chance that the effect will be above this level).

2. Suppose I now tell you that the effect is below this level; tell me your new 50% level: this represents the level at which you think there is a 25% chance that the effect will be below it.

3. What is the smallest credible level for the effect? By this, I mean what level do you think it is one hundred to one against the effect being below this level?

4. Can you be certain that the effect will not be below any level (you would bet your life on it)? If so, what is this level?

The last three questions were repeated for high levels.

As we saw in Chap. 10 it is not difficult to find a probability distribution which best fits these judgements. Looking at the fifteen responses, it seemed reasonable to summarise the different views as being in one of three camps: the optimists, the pessimists or the disinterested. These three points of view were described by three base rate distributions (Fig. 14.18).

Fig. 14.18 Base rate distributions model different points of view

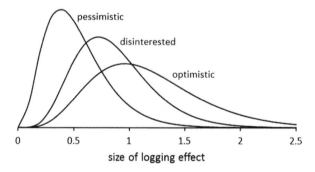

The optimistic view, which might be typical of loggers, has a peak value of about 1.0, the value describing no-change; logging has no effect on the abundance of the birds.

On the other hand, the pessimistic view, which a conservationist might hold, is very much skewed towards lower values indicating a belief that logging will cause a reduction in bird numbers. The most likely value is about 0.4, a sixty percent reduction.

The third view is between these two. It was called disinterested in the sense that someone holding that view had no direct stake or keen interest in the effects of logging; an ordinary member of the public.

The spread of each of these base rate distributions shows that for each person and within each group there is some uncertainty about just how big the effect of logging might be. Even a pessimist believes with probability seven percent that logging might increase abundance.

With each of these base rate distributions the Bayes analysis was repeated (Fig. 14.19). After revising belief in light of the evidence of the survey data the three distributions are very similar (and similar to Fig. 14.17 too): there has been a convergence of opinion given the data, a familiar result.

Fig. 14.19 Data leads to convergence

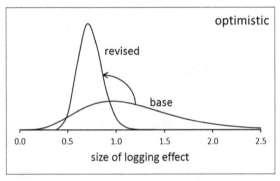

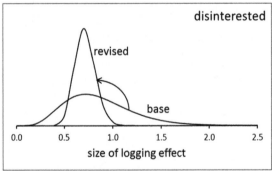

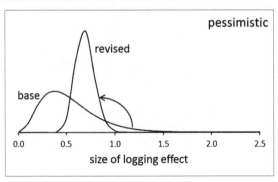

Decisions about logging are important commercially and ecologically. They will be informed by these estimates of the impact on birds and mammals. Being able to show that the views of interested parties converge given the survey results should help in demonstrating accountability in the decision making.

~ • ~

Bayesian methods are increasingly used in ecology [8][10] for the two reasons illustrated by the cases in this chapter. First, models of ecological processes can be complex and so likelihoods difficult to handle in any other way. Second, base rates allow judgement to be used either, as here, because there exist competing viewpoints or because, when data are scarce, base rates provide a way to import results from elsewhere.

~~~ ••• ~~~

References

1. Food and Agriculture Organization of the United Nations (2015) Global forest resources assessment 2015. FAO, Rome. http://www.fao.org/3/a-i4808e.pdf. Accessed 24 Sept 2017
2. Food and Agriculture Organization of the United Nations (2017) Forest products statistics http://www.fao.org/forestry/statistics/80938/en/. Accessed 24 Sept 2017
3. DeBald PS, Mendel JJ (1976) Standard evaluation tools III. Composite volume and value tables for hardwood pulpwood. US Department of Agriculture, Forest Service Research Paper NE-338. Forest Service US Department of Agriculture Northeastern Forest Experiment Station, Darby, PA
4. Platt WJ, Evans GW, Rathbun SL (1988) The population dynamics of a long-lived conifer (*pinus palustris*). Am Nat 131(4):491–525
5. Beckage B, Joseph L, Belisle P, Wolfson DB, Platt WJ (2007) Bayesian change-point analysis in ecology. New Phytol 174(2):456–467
6. Thomson JR, Kimmerer WJ, Brown LR, Newman KB, MacNally R, Bennett WA, Feyrer F, Fleishman E (2010) Bayesian change point analysis of abundance trends for pelagic fishes in the upper San Francisco estuary. Ecol Appl 20(5):1431–1448
7. Crome FHJ, Thomas MR, Moore LA (1996) A novel Bayesian approach to assessing impacts of rain forest logging. Ecol Appl 6(4):1104–1123
8. King R, Morgan BJT, Gimenez SP, Brooks SP (2010) Bayesian analysis for population ecology. Chapman & Hall, Boca Raton

---

[10]For a flavour of earlier discussions making the case for Bayesian methods see the papers in the journal Ecological Applications for November 1996 (vol. 6, no. 4).

Chapter 15
Radiocarbon Dating

Archaeologists need to know the age of what they find; human and animal bones, for instance. One of the main methods to help them is radiocarbon dating. The level of radioactivity in an organism (you and me included) is constant while we are alive but declines at a known and almost constant rate after death. Measuring the reduced level of radioactivity of archaeological finds provides an estimate of their age. These estimates are subject to some error. Knowing the magnitude and shape of the distribution of errors means that likelihoods can be constructed and so the age of the archaeological material estimated from the level of radioactivity.

Quite often an archaeologist will have some other information. For example, the style of decoration used on pottery fragments found at the dig might give a good idea of which period or century a settlement was established. This base rate estimate is made independently of the radiocarbon dating and so Bayes' Rule is a good way of combining the two.

~ • ~

We often think of carbon dioxide as a bad thing, a product of burning fossil fuels in cars and factories which then rises in the atmosphere to increase global warming. All of which is true, but carbon dioxide is also the vital start of the radiocarbon dating process.

All living things are in part made of carbon. The carbon cycle describes how this happens. Carbon dioxide exists in the atmosphere. Photosynthesis is the process by which plants convert energy from the sun into chemical energy and then into the physical matter of the plant itself. During this process atmospheric carbon becomes part of the plant. Animals, ourselves included, eat plants. Some animals also eat other animals who themselves have eaten plants, of course. And so both plants and animals are made, in small part, of carbon. The decomposition which follows death releases carbon dioxide back into the atmosphere by a process called respiration. And so the cycle is complete.

All chemical elements can exist in a number of forms, called isotopes. Carbon exists in three forms, ^{12}C, ^{13}C and ^{14}C. The third of these, called carbon fourteen or radiocarbon, is radioactive, the radioactivity being the result of bombardment by cosmic rays in the atmosphere about fifteen km above the earth. Although ^{14}C

A. Jessop, *Let the Evidence Speak*, https://doi.org/10.1007/978-3-319-71392-2_15

constitutes only about a millionth part of a millionth part of carbon that is enough. The point is that ^{14}C is unstable. The level of radioactivity in plants and animals declines but, for so long as the plant or animal is alive, carbon is replenished and a stable state is maintained. When the plant or animal dies this replenishment stops and so the level of radioactivity decreases. Radiocarbon dating is based on the rate at which this happens.

During the Second World War Willard Libby worked on the Manhattan Project, which resulted in the bomb that destroyed Hiroshima. After the war he went as a Professor of Chemistry to the Institute for Nuclear Studies and Department of Chemistry at the University of Chicago. In 1946 he wrote a letter to the editor of the journal Physical Review in which he set out the basis for radiocarbon dating [1], further described at book length in 1952 [2]. In 1960 he received the Nobel Prize for Chemistry.

Libby argued that the rate of decline of radioactivity could be described by a simple curve. Given the initial level of 15.3 dpm/gC (disintegrations per minute per gramme of carbon) the level of ^{14}C radioactivity in archaeological remains is measured and their age found.

The rate of decline is usually described by the element's half-life, the time taken for the level of radioactivity to be halved. Libby estimated the half-life of ^{14}C to be 5568 years. After this time radioactivity would be half the original value at the time of death. After another 5568 years it would halve again to a quarter of the original value and so on (Fig. 15.1).

Fig. 15.1 Libby's model of radioactive decay

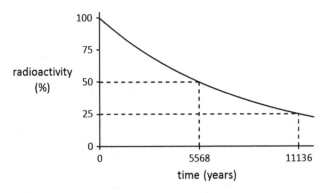

In his analysis Libby made several assumptions, for instance that the amount of ^{14}C in the atmosphere has been constant, which have proven not to be true. The half-life of ^{14}C is now established as 5730 years. The general shape of the model follows Libby's curve but is not nicely smooth due to irregularities introduced by modifying his simplifying assumptions. The radiocarbon age of an archaeological sample, now found using Accelerator Mass Spectrometry (AMS), needs to be calibrated against known true age. The AMS result can then be used to estimate the age of the archaeological sample. This calibration requires some samples of known age. These are provided primarily by samples of wood which can be dated accurately by an analysis of tree rings.

Archaeologists count back from now and so measure years on a scale Before Present, BP. The starting point on this scale is 1950, which is when radiocarbon dating began, so the year 1850 is 100BP. On the graph in Fig. 15.2, the vertical axis shows the date estimated by the radiocarbon dating, called ^{14}C BP or radiocarbon years. The horizontal scale shows calendar years, cal BP (alternatively, calibrated years). The result is a calibration curve.

Fig. 15.2 Calibration curve

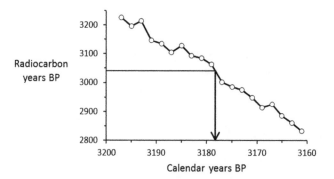

The irregularities (known as "wiggles") are modelled as a series of straight lines. From the date estimated by a laboratory, the radiocarbon years, the best estimate of the calendar age can be found. In the graph, a radiocarbon age of 3040BP gives a calendar age of 3178BP or 1228BC.

The estimate is not precise. All samples with calendar age 3178 should have the same radiocarbon date but this is not so. Differences in sampling, variation in laboratory practice and so on mean that rather than a nice sharp line we have a band, the width of which depends on these estimation errors. The date given by the graph is the average value.

Because of these variations a laboratory will usually give an estimate as a range so that instead of 3040 there will also be an estimation error, typically a Normal distribution with standard deviation about 60 years, so you might see 3040 ± 60. A useful way to think about this is that there is a 95% chance that the true radiocarbon age is in an interval of twice the standard deviation, so a lower limit of 3040 − 120 = 2920 and an upper limit of 3160. These translate to calendar years of 3169 to 3192 (Fig. 15.3).

Fig. 15.3 Estimate of calendar age showing error interval

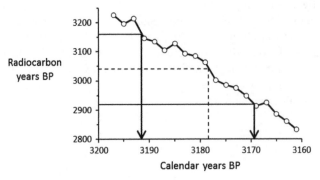

This is called the intercept method of finding calendar dates from radiocarbon measurements and has been widely used.

~ • ~

Although of limited archaeological interest the Turin Shroud is, for other reasons, of great interest to many.

The linen cloth measuring about 434 cm by 109 cm shows an image of a man who seems to have been whipped and crucified (Fig. 15.4).

Fig. 15.4 Image on the Turin Shroud[1]

[1]Image, Secondo Pia. Wikimedia Commons https://commons.wikimedia.org/wiki/File:Turin_plasch.jpg. Accessed 24 September 2017.

It was first shown at Lirey, near Troyes, in France probably between 1355 and 1453. It was the property of the mediaeval knight Geoffry de Charny who was killed in 1356, during the Hundred Years War. He left no account of how he came by the cloth.

The cloth was believed to be the burial shroud of Christ and the image the face of Christ. As such it was, and for some remains, an object of veneration. After some travels it was placed in the Royal Chapel in Turin Cathedral in 1694.

The history of the cloth before Geoffry obtained it is unknown but it was believed to have been kept in Constantinople since the time of Christ. There were many competing theories about the shroud and its origins. Surely, radiocarbon dating could help by at least determining the age of the cloth. Although this possibility had been obvious since the early days of dating the methods then available would have required a sample for analysis of about 500cm^2, much too large to be cut from the shroud. The development of AMS technology in the 1970s meant that radiocarbon dating now only needed samples of a few square centimetres, about the size of a postage stamp. And so on 21 April 1988 in the Sacristy of Turin Cathedral a group of scientists, in the presence of the Archbishop of Turin, Cardinal Anastasio Ballestrero, took samples of the cloth. These samples were sent to three laboratories—in Oxford, Arizona and Zurich—for dating to be carried out [3].

Figure 15.5 shows the analysis of one of the samples. As with the earlier graph the 95% interval estimate of the radiocarbon age is shown on the vertical axis. These limits are projected until they reach the calibration curve from which they are projected down to give a corresponding estimate of the calendar age of the cloth. In this case the jaggedness of the calibration curve gives a calendar age estimate in two parts, two ranges, from 1262 to 1312 and from 1353 to 1384.

Fig. 15.5 Radiocarbon dating of the Turin Shroud[2]

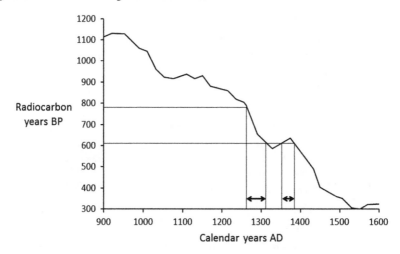

Radiocarbon years BP

Calendar years AD

---

[2]Redrawn from Damon et al. [3] Fig. 2, p. 614).

From the analysis of all results it was concluded that there is a ninety-five percent probability that the shroud dates from between 1260 and 1390. This estimate includes the date of its first exhibition at Lirey.

You will not be surprised to know that there were disbelievers (of this result) but none of the objections were credible [4]. The cloth is mediaeval.

~ • ~

The intercept method is a very clear way of giving calendar age dates. Given upper and lower limits of the radiocarbon determination corresponding range estimates of the calendar date convey uncertainty. But from a Bayesian point of view the relation is the wrong way round.

The whole point of the analysis is to use the radiocarbon date as *evidence* to help us decide how much we should believe in *alternative* values of the calendar age of the specimen. The specimen is old because of its calendar age. The radiocarbon age is as it is because of the calendar age, not vice versa.

Because of measurement error specimens with the same calendar age will generally have different radiocarbon ages. It is this distribution of radiocarbon dates which is the likelihood distribution.

Before we got diverted by the Turin Shroud we saw in an illustration that using the calibration curve a radiocarbon date of 3040 corresponded to a calendar date of 3178. To help in making their report the laboratory used the data on experimental error to give

probability that calendar age is 3178 GIVEN radiocarbon age is 3040

but for the Bayes Grid we need likelihoods which give

probability that radiocarbon age is 3040 GIVEN calendar age is 3178

In the previous chapter we saw how the heights of the Normal distribution gave relative likelihoods (Fig. 14.9). Figure 15.6 shows a likelihood distribution for radiocarbon dating.

Fig. 15.6 Likelihood distribution for radiocarbon dating

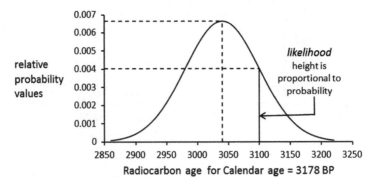

The Bayes Grid for this analysis is shown in Fig. 15.7.

Fig. 15.7 Bayes Grid for radiocarbon dating

alternatives:	evidence: radiocarbon year					base
calendar year	...	3040	...	3100	...	rate
3178	...	0.0066	...	0.0040	...	?
3179		←	likelihood	→		?

Years are given here as whole numbers but time isn't like that. We could have 3100.84. . . and so we'll have distributions as curves rather than tables. This is just what happened when we looked at the percentage answering Yes in the referendum poll in Chap. 6.

In practice the unevenness of the calibration function (the wiggles) means that the estimate of the calendar age of the sample will be similarly uneven, something like Fig. 15.8.

Fig. 15.8 Typical estimate of calendar age

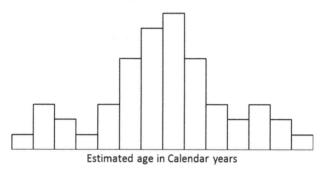

Estimated age in Calendar years

What of base rates?

Archaeologists often have base rate information they wish to use. It is sometimes in a different form to that which we have seen so far. The calculations can be difficult. Fortunately, there is software to help, such as the OxCal program of Oxford Radiocarbon Accelerator Unit at the University of Oxford.

One problem which often faces archaeologists is sequencing. As well as evidence in the form of samples of wood or bones which can be radiocarbon dated there may be other good reasons for believing that some samples are older than others. This base rate information acts as a constraint on the estimates of age.

~ • ~

Ecuador is located in the north west of South America. To the west is the Pacific, to the north Columbia and to the east and south is Peru. The Jama River is in the northern Manabí province and is known as the centre of the Jama-Coaque culture which lasted about 2000 years from 355 BC to 1532 AD [5]. As at the mid-1990s the time and space characteristics of the archaeology of this area were not well understood. James Zeidler, Caitlin Buck and Clifford Litton set out to use the Bayesian approach to radiocarbon dating to provide a better chronology for the Jama valley [6]. As they wrote, "Chronology construction is one of the most fundamental tasks in archaeological research, yet one that is never finished."

A very general sequence of the cultures of the Jama valley had been established in the 1950s, based primarily on a little fieldwork at a few sites on the coastal strip. This enabled a general sequencing (ordering) but no chronology (dating). A periodization was developed in 1957 by the Ecuadorian archaeologist Emilio Estrada. This was taken up by Betty Meggers of the Smithsonian Institution [7]. Figure 15.9 shows the scheme, proposed as descriptive of the archaeological history of Ecuador in general.

Fig. 15.9 The sequencing of Estrada and Meggers[3]

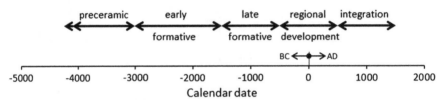

An important source of information for archaeologists is pottery and ceramics, changes in style often indicating changes in culture. While the cultures cannot be examined some of their artefacts can. For example, the ceramic bottle in Fig. 15.10 is from the Chorrera culture in Ecuador dating from the ninth century BC to the second century BC.

---

[3]This diagram uses information from Meggers [7].

Fig. 15.10 Chorrera spout and bridge bottle[4]

Because of the central role of pottery in establishing a sequence of periods, the work of the archaeologists was restricted to the four periods following the preceramic: Early and Late Formative, Regional Development and Integration.

~ • ~

Research over the ten to twenty years following the work of Estrada and Meggers continued mostly in coastal regions and so did not include northern Manabí province. There was much illegal looting of archaeological sites. Some of these looted artefacts were held by local museums and it was these that were used subsequently for study. But these artefacts were no longer reliably linked to the places from which they had been taken and so were of only limited use, if any, in establishing a chronology for particular areas.

In Fig. 15.9 it is assumed that one period followed another in a pattern of continuous development, but in fact there had been breaks in the sequence, mostly due to volcanic eruptions and the Spanish conquest.

---

[4]Chorrera bottle. Image, Metropolitan Museum of Art. Wikimedia Commons. https://commons.wikimedia.org/wiki/File:Spout_and_Bridge_Vessel_MET_1988.117.5_a.jpg. Accessed 24 September 2017.

Three main cultural phases were identified for the Jama valley. The earliest was the Valdivia culture. Common in the western lowlands and coastal region, it saw the start of a settled village life based on horticulture. The Valdivia was itself subdivided into phases the first seven of which were in the preceramic period. The eighth and final phase was marked by the establishment of larger inland centres of ceremonial significance and with satellite settlements.

The Chorrera culture was geographically widespread, from the coastal lowlands up onto the Andean highlands. The Chorrera was mainly distinguished by its ceramic vessels (one of which is shown above), particularly effigy bottles and other figures.

Next, and most long-lasting, was the Jama-Coaque culture. This comprised a set of chiefdoms with strongly hierarchical social and political organisation. Position in the hierarchy was signalled by dress and body ornamentation.

The Jama-Coaque culture was ended by the Spanish conquest, with a consequent decline in both the quality of ceramic art and, thanks to imported disease, of the human population too. The next phase, sometimes named Campace after people who lived a little south of the Jama River valley, broadly followed the Jama-Coaque tradition.

Within each culture there existed a characteristic ceramic style, in this area labelled Piquigua, Tabuchila, and Muchique. These styles can sometimes be further subdivided into periods of their own. There are two Piquigua periods and five Muchique. This more detailed periodization was used by the archaeologists.

The Spanish conquest was not the only source of discontinuity. Ecuador has volcanoes (twenty-five are currently listed) and when they erupt they cause effects so severe that settlements may be wiped out and new ones, after a time, take their place: a break in the pattern. The damage comes not just from the molten lava, images of which you will have seen, but also from the volcanic ash ejected during an eruption, which is called tephra. The ash may travel a long way before landing, hundreds of kilometres perhaps. The layers of tephra which are deposited are good indicators of critical points in a chronology. Analysis of granularity and composition help to locate the source of the eruption. The Jama valley suffered three such volcanic disruptions; one split the Piquiga period and one occurred either side of the first Muchique period. There was also a non-volcanic hiatus between the Piquiga and Tabuchila periods.

There were only a few samples suitable for radiocarbon dating for the two Piquiga phases and so they were treated as one. The resulting seven phases used by Zeidler, Buck and Litton are shown in Table 15.1.

phase	culture	ceramic phase	break	period
7	Campace	Muchique 5		Colonial
			Spanish conquest	
6	Jama-Coaque II	Muchique 4		Integration
5	Jama-Coaque II	Muchique 3		Integration
4	Jama-Coaque II	Muchique 2		Integration
			Tephra III	
3	Jama-Coaque I	Muchique 1		Regional development
			Tephra II	
2	Chorrera	Tabuchila		Late formative
			hiatus	
1	Valdivia 8	Piquiga (late)		Early formative
			Tephra I	
	Valdivia 8	Piquiga (early)		Early formative

Table 15.1 Seven phases of settlement[5]

Using the field experience of the archaeologists and the opinions of others expert in the archaeology of Ecuador a number of points seemed clear. The four oldest phases, from Piquiga to Muchique 2, followed one from the other. There were thought to be gaps between them. The two most recent periods, Muchique 4 and Muchique 5, abutted each other so that the end of one was coincident with the start of the other. Muchique 3 was thought to overlap its neighbours. This consensus is shown in Fig. 15.11.

Fig. 15.11 Sequencing of the seven phases[6]

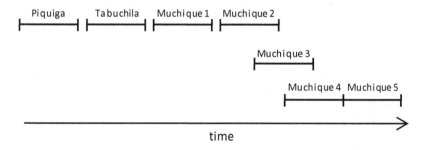

This sequencing was fine, but what of the dates of these phases?

Each of the seven phases is described by two values, a start date and an end date. These are the fourteen parameters we wish to estimate. Just as in other applications base rate information has described some prior knowledge or context. In this case

[5]Zeidler, Buck and Litton [6], Table 1, p. 165.
[6]Based on Zeidler, Buck and Litton [6] Fig. 3, p. 171.

the sequencing information provides that knowledge but as a series of inequalities rather than as a probability distribution. There were seven obvious constraints to ensure that no phase ended before it started. Counting in years BP, so that earlier events have larger numbered dates,

> *start* Piquiga is greater than *end* Piquiga

and so on for all seven phases. The three gaps between the four earliest phases were ensured by

> *end* Piquiga is greater than or equal to *start* Tabuchila
> *end* Tabuchila is greater than or equal to *start* Muchique 1
> *end* Muchique 1 is greater than or equal to *start* Muchique 2

These three constraints were "greater than or equal to" and not just "greater than" to allow for the possibility that the analysis might give a gap width of zero, and so provoke second thoughts about the effects of the disruptions.

For the three phases Muchique 2, Muchique 3 and Muchique 4 it was important that they both started and ended in that order, as shown in the diagram and so,

> *start* Muchique 2 is greater than *start* Muchique 3
> *start* Muchique 3 is greater than *start* Muchique 4

and

> *end* Muchique 2 is greater than *end* Muchique 3
> *end* Muchique 3 is greater than *end* Muchique 4

Finally, to ensure that Muchique 4 and Muchique 5 abut each other

> *end* Muchique 4 is equal to *start* Muchique 5

~ • ~

These conditions set the relative ordering of the fourteen parameters. The dates were estimated using a Bayes model with, as evidence, thirty-six samples of charcoal from sixteen sites. Care was taken that wherever possible these samples were from well defined features such as pits and hearths. This was to ensure the closest correlation between the results of the carbon dating and actual locations in the stratigraphic record. There was a thirty-seventh sample which had previously been used by Estrada and was included as corroborating evidence.

~ • ~

The likelihood distribution rested on three assumptions. First, that for any calendar date the distribution of radiocarbon dates followed a Normal distribution, as described earlier. Second, that for each sample and within each phase, the calendar date followed a uniform (flat) distribution. This is just like the equal base rate assumptions seen in many of the earlier applications. Third, the evidence from different samples are assumed to be statistically independent; another familiar assumption.[7]

The results of the analysis are probability distributions for each of the fourteen dates. Figure 15.12 shows the intervals within which ninety-five percent of each distribution falls (horizontal line) and the most likely value, the mode or peak of each distribution (circle). The dates are calendar years.

Fig. 15.12 Estimated dates for the seven phases[8]

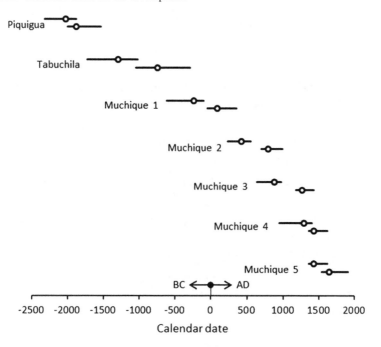

The top bar of each pair shows the probability estimate of the start date and the bottom bar the end date.

~ • ~

[7]A mathematical description of the model is given in Buck et al. [8]. You may also be interested in the implementation of this model in the winBugs software by Andrew Millard of Durham University   at   http://community.dur.ac.uk/a.r.millard/Bucketal9_5.txt.   Accessed 25 September 2017.
[8]Based on Zeidler, Buck and Litton [6] Table 4, p.174.

One of the benefits of having these probabilistic estimates is that the possibility of overlaps can be evaluated, particularly as it effects Muchique phases 2, 3 and 4. Zeidler, Buck and Litton found that

the probability that Muchique 2 completely precedes Muchique 3 is about 60%

it is almost certain that Muchique 2 completely precedes Muchique 4

the probability that Muchique 3 completely precedes Muchique 4 is almost 50%

Based on these and other evaluations they therefore felt it reasonable to conclude that Muchique 2 precedes Muchique 4 but that the existence of the other overlaps between these three phases remain uncertain.

~ • ~

Using Bayes' Rule for the interpretation of archaeological and radiocarbon data enabled both phasing and dating information to be brought together in one unifying framework

likelihood   enabled an estimate of the calendar date in probability form to be found from the evidence of the radiocarbon date

base rates   enabled what was known about the sequencing of the seven phases to be used as part of the date determination

There has, of course, been much subsequent work, both in the archaeology of Ecuador [9] and in the application of Bayesian analysis to archaeological problems,[9] including in the estimation of calibration curves [10].

~~~ ••• ~~~

References

1. Libby WF (1946) Atmospheric helium three and radiocarbon from cosmic radiation. Phys Rev 69(11–12):671–672
2. Libby WF (1952) Radiocarbon dating. The University of Chicago Press, Chicago
3. Damon PE, Donahue DJ, Gore BH, Hatheway AL, Jull AJT, Linick TW, Sercel PJ, Toolin LJ, Bronk CR, Hall ET, Hedges REM, Housley R, Law IA, Perry C, Bonani G, Trumbore S, Woelfli W, Ambers JC, Bowman SGE, Leese MN, Tite MS (1989) Radiocarbon dating of the Shroud of Turin. Nature 337:611–615
4. Gove HE (1990) Dating the Turin Shroud—an assessment. Radiocarbon 32(1):87–92
5. Masucci MA (2008) Early regional polities of coastal Ecuador, Ch. 25. In Silverman H, Isbell WH (eds) Handbook of South American Archaeology. Springer, New York, pp 489–503

---

[9]In 2015 the journal World Archaeology had a special issue, 47(4), on Bayesian methods in archaeology.

6. Zeidler J, Buck CE, Litton C (1998) Integration of archaeological phase information and radiocarbon results from the Jama River valley, Ecuador: a Bayesian approach. Lat Am Antiq 9(2):160–179
7. Meggers BJ (1966) Ecuador. Thames and Hudson, London
8. Buck CE, Cavanagh WG, Litton CD (1996) Bayesian approach to interpreting archaeological data. Wiley, Chichester, pp 226–233
9. Silverman H, Isbell WH (eds) (2008) Handbook of South American Archaeology, Chapters 24–27. Springer, New York, pp 459–453
10. Blackwell PG, Buck CE (2008) Estimating radiocarbon calibration curves. Bayesian Anal 3 (2):225–248

Chapter 16
Bayes and the Law

Say "evidence" and crime and the law are likely to be the first things that come to mind. You would expect Bayes thinking to be particularly useful both to police in sifting evidence and to lawyers and others in court who have to use that evidence. We saw some of this in the very first chapter of this book. Now we'll look in a little more detail at some problems in the courtroom.

Unlike other applications in this section this is not a straightforward success story. The history of the use of statistics and probability, and then Bayes' Rule, by witnesses and lawyers has not always been a happy, or even a competent, experience. But Bayes arguments have been important in making clear errors in the use of evidence. Getting something useful done to improve the situation is another thing entirely.

~ • ~

A trial involves a number of people; lawyers, judges, eyewitnesses, expert witnesses, and last, but by no means least, the jury (If there is one. South Africa, for instance, does not rely on juries).

All are there because of the need to make a decision based on evidence. Much evidence is subject to some degree of uncertainty, for example identifying that the shoe print found at the crime scene was made by a shoe owned by the accused. It would be natural to talk about the uncertainties of this sort of evidence by giving probabilities but, as we saw in Chap. 10, we are not particularly well equipped to use this language. Expert witnesses, particularly forensic experts, can in most cases give probability estimates but what do the lawyers, the judge and, crucially, the jury make of it all? There are a number of well known cases which show the difficulties [1].

~ • ~

Steve and Sally Clark were two young lawyers living in Chester, in the North of England. They had two children, both of whom died soon after birth. Christopher died in 1996 aged ten weeks. Harry died in 1998 aged eight weeks.

Because such tragic events are, thankfully, unusual we look for reasons. It is human nature to do so. Both deaths could have been accidental as the result of Sudden Infant Death Syndrome, SIDS, also called cot death. One such death is a tragedy.

© Springer International Publishing AG, part of Springer Nature 2018   189
A. Jessop, *Let the Evidence Speak*, https://doi.org/10.1007/978-3-319-71392-2_16

Perhaps two cot deaths are so rare that a more likely explanation may be that the deaths were not accidental at all and that both boys had been killed by their mother.

In 1999 Sally Clark was tried for murder.

A key witness for the prosecution was the leading paediatrician Sir Roy Meadow. He gave expert testimony that there was only a 1 in 73 million chance of two cot deaths in the same family. This was taken to be so very small that it provided overwhelming support for the charge of murder. Sally Clark was found guilty and sent to prison.

In 2000 Sally Clark appealed on two grounds. First, that in calculating his figure of 1 in 73 million Meadow had made an assumption of statistical independence and that this was not valid. Second, that the jury may well have misinterpreted Meadow's figure to be the probability that she was innocent, which it was not.

$$\sim \bullet \sim$$

Meadow used figures taken from the Confidential Enquiry for Stillbirths and Deaths in Infancy, CESDI, which was based on data from five regions of England for the period 1993–1996. This gave the chance of cot death as 1 in 1300. For an affluent non-smoking family with the mother over twenty-six, which the Clarks were, this chance fell to 1 in 8500. Meadow took this figure and, on the assumption that cot deaths were independent of each other, simply multiplied to give a probability of $(1/8500) \times (1/8500)$ or 1 in 73 million. (Lawyers prefer to express uncertainty using odds such as these rather than as probabilities, finding them more comprehensible, especially for very small probabilities.)

What causes cot deaths? Since siblings are genetically similar, the possibility that there is some genetic influence on cot deaths is important. In fact there is such a genetic predisposition. Once we know that the first death was caused in this way the chance that the second death was also a cot death is higher, between 1 in 60 and 1 in 130. Say 1 in 100. Using this figure the probability of two cot deaths is $(1/8500) \times (1/100)$ which is 1 in 850,000, a very long way from 1 in 73 million. Meadow's error was to overlook this possibility and to assume that the two deaths were statistically independent.

Even so, a 1 in 850,000 chance, just 0.00012%, might still be taken as small enough that a jury will think guilt all but certain. But they would be wrong. To see why use a little Bayes thinking (Fig. 16.1).

Fig. 16.1 Bayes Grid asks questions about presentation of evidence

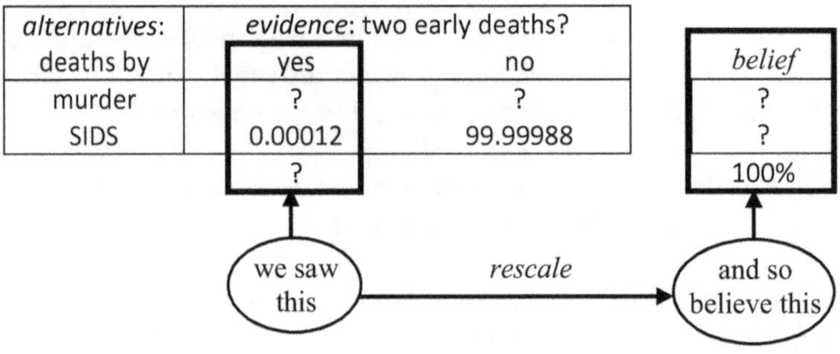

Meadow had incorrectly calculated the likelihood of two cot deaths. But even correcting for that the jury might easily mistake what they heard. They may confuse the probability of two deaths given that they were accidental and Sally Clark innocent, which Meadow gave, with the probability that Sally Clark was innocent given that the two deaths had happened, which he had not. If they thought they had heard Meadow give the second, and being impressed by just how very small that probability was, they might unjustifiably conclude that Sally Clark had murdered her sons.

The Appeal Court accepted the first objection, that the assumption of independence was unjustified, but rejected the second. The judges said that it was "stating the obvious" that the two probabilities were not the same. The appeal was dismissed.

~ • ~

It was at this point that statisticians became alarmed. Professor Peter Green, in his role as President of the Royal Statistical Society, RSS, wrote to the Lord Chancellor, England's senior law officer, setting out objections to Meadow's evidence.[1]

In addition, Professor Ray Hill of the University of Salford looked at the CESDI report and gave a better estimate of the probabilities of the cot deaths [2].

In 2003 there was a second appeal. The defence first presented evidence of a blood test obtained by the original pathologist that Harry was suffering from a respiratory infection at the time of his death. The Clark's first child almost certainly died of natural causes. The blood test report, from Macclesfield Hospital, had not been made available at the original trial or the subsequent appeal. On the second day of this second appeal, following the evidence of the blood test, the prosecution told the judges that it no longer sought to uphold the convictions. In its judgement the Court of Appeal criticised Meadow's original evidence as "manifestly wrong" and "grossly misleading".

Sally Clark was freed.

After her release Sally Clark found it hard to come to terms with what had happened. She died in 2007 from alcohol poisoning.

~ • ~

Although Sally Clark's acquittal was, in the event, due to the failure to make known the results of the blood test her case raised important statistical issues. Let's go back to the original trial.

---

[1]The Royal Statistical Society. Letter from the President to the Lord Chancellor regarding the use of statistical evidence in court cases, January 23rd 2002. http://www.rss.org.uk/Images/PDF/influenc ing-change/rss-use-statistical-evidence-court-cases-2002.pdf. Accessed 26 September 2017.

Likelihoods must be as accurate as we can make them or all else is wrong too. The RSS letter warned that the Clarks had not been characterised independently of the way in which the data had been described; affluent, non-smoking, mother over twenty-six. They had been put in an already defined subset of the CESDI data. The RSS said that it would have been better to use the chance of 1 in 1303 characteristic of the whole population. Had that been done the figure of 1 in 73 million would have become 1 in 600,000. In addition, recognising the genetic predisposition to cot death and so using the greater probability of the second death gives an increased likelihood of $(1/1300) \times (1/100)$ which is 1 in 130,000, or 0.00077%. This is a very long way from Meadow's figure of 1 in 73 million.

The Bayes Grid also makes it clear that something is missing. We have the likelihood of two early deaths if the cause was SIDS but we have no similar likelihood if the cause was murder. It is fundamental that it is the *relative* values of likelihoods that is important: we need both.

The Home Office estimates that fewer than 30 babies are killed by their mother each year. There are 650,000 births a year and so the probability that a baby is murdered is no bigger than 30/650,000. Double murders are much less common. Helen Joyce, then Editor of Cambridge University's *Plus* magazine, argued for a probability of double murder smaller by a factor of ten [3] to give a probability of 0.00046% that both Christopher and Harry were murdered. This is a conservative estimate.

With both likelihoods we can complete the Bayes Grid (Fig. 16.2).

Fig. 16.2 Evaluation of evidence for Sally Clark

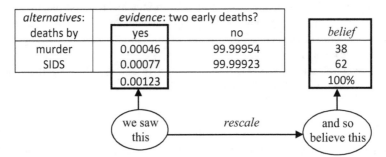

The conclusion justified by the data is that there is only a 38% chance that the two boys were murdered.

Sally Clark was most probably innocent. Even though some approximations were used to get the likelihoods (Helen Joyce's use of the factor of ten to account for double murder, for instance) it is unlikely that any plausible alternative assumptions would increase the 38% so much that the jury could reasonably conclude that Sally Clark was guilty "beyond all reasonable doubt", which is what English criminal law requires.

~ • ~

Meadow had made a quite basic error when he assumed that two cot deaths were statistically independent and so just squared the probability of one death. It should have been obvious to an expert, which Meadow was, that this strong assumption needed to be supported by evidence. Not all expert witnesses are also expert in statistics.

In 2005 The General Medical Council, GMC, found that Sir Roy Meadow's misuse of statistics in the Sally Clark trial amounted to serious professional misconduct and struck him from the medical register which meant that he could no longer practice. Meadow appealed and in the following year the High Court found in his favour, concluding that the GMC had properly been critical of his performance as an expert witness but wrong in concluding that this constituted serious professional misconduct. He was reinstated and could practice again, though he retired soon after.

~ • ~

In the trial of Sally Clark there was never any suggestion that her children could have been murdered by anyone else: either Sally Clark murdered her children or they died of SIDS. This is not always the case. For example, if the only evidence is the imprint of a shoe of particular type and size and the accused has such a shoe, the defence may reasonably say that there must surely be a great many other people in the city or the country also possessing shoes of just this type and size. To disregard this base rate information is the same error that a doctor would make if the prevalence of a disease or illness in the population were ignored when making a diagnosis (Chap. 8).

A careless but eager prosecutor may make two potentially misleading mistakes in presenting evidence to a jury. First, mistaking just what probabilities mean, as in the Sally Clark case, and confusing the probability that the deaths occurred given that Clark was innocent with the probability that she was innocent given the deaths. This is called the error of the transposed conditional. Second, base rates may be ignored. William Thompson and Edward Schumann named such false reasoning *The Prosecutor's Fallacy* [4]. It has become common to use The Prosecutor's Fallacy to mean only the first error, the transposed conditional.

Evidence based on DNA analysis is increasingly used in trials (not just in those television police shows). Surely the precision of DNA profiling will give such good evidence that it will swamp any of these concerns in just the same way that an estimate based on a large survey will swamp any prior belief, as we saw in Chap. 14. But will it? Suppose that, based on a DNA sample, the probability that it was from the accused is one in a million. This sounds convincing. Suppose again that it is known that the crime was committed by a man older than about twenty but not more than mid-sixties. There are about twenty million men in this age group in the UK any of whom *may* provide a match to the DNA found at the crime scene.

What is the Court to make of this?

~ • ~

In November 1989 a 65 year old woman was sexually assaulted in her home at Ashton-under-Lyne in Greater Manchester. A year later Alan James Doheny was tried at Manchester Crown Court where he was convicted of this assault and sentenced to eight years imprisonment. In 1991 this was increased on appeal to twelve years.

The prosecution case relied heavily on the evidence of an expert witness that DNA found at the crime scene matched Doheny's DNA.

DNA consists of two intertwined strands (the double helix), one from the mother and one from the father. Think of each strand as a code. The code has four elements or chemical bases: A (adenine), G (guanine), C (cytosine), and T (thymine) and might look something like this GTAATCCGA...

The code for human DNA is three billion characters long. 99.9% of our DNA is common to us all. The other 0.1% is highly variable between individuals. It this 0.1%, or part of it, which is used.

A strand is divided into a number of sections of code. These are called locations and are given names such as D21 or D7S280. It is possible that at some locations short sections of code are repeated. For example in the section below the four characters GATA (shown in capitals) are repeated four times

gtatcaactattgtacccggaacg**GATAGATAGATAGATA**ctgttacatgggagcaga

If at the same location on the other strand GATA were repeated six times the profile at this location would be given as the pair 4,6.

These short repeating sequences such as GATA are called Short Tandem Repeats (STRs) or markers. STRs are usually of length four but other lengths are sometimes used. A DNA profile using four locations might look like that shown in Fig. 16.3.

Fig. 16.3 Hypothetical DNA profile

location:	D21	THO1	D19S433	D7S280
STR pair:	16,14	9,10	27,30	14,15

It is the job of the forensic scientist to compare two profiles—one from the crime scene and one from the suspect—and decide whether there is a match or not.

The process of analysis causes repetitions of the STRs around their original locations (loci) which has the effect of blurring the profile. These repetitions are called stutters and the blurred location is a stutter band.

The sample taken from a suspect by mouth swab should be well defined. The sample taken at the crime scene may be degraded due to the age of the sample, contamination at the crime scene, or attempts by the perpetrator to remove traces of fluids or other deposits from which DNA might be found. In addition, laboratory procedures are sometimes imperfect.

To reduce the chance of error when comparing two DNA samples more than one marker (location) is tested, typically thirteen nowadays but fewer at the time Doheny was arrested. If the crime scene sample is degraded fewer markers may be available for comparison.

If we can be sure that two profiles do not match then we can be sure they are from different sources. If they do match it is highly likely that they are from the same source. But since not all possible markers are examined, there is always a small chance that this profile is shared with someone else.

The analysis enables an estimate of the proportion of a population, or group within that population, with the same profile as that found at the crime scene. This is the match probability or random occurrence ratio, also sometimes called the random match probability.[2]

~ • ~

In the trial of Alan Doheny the prosecution case relied mostly on DNA evidence. An expert witness, a Mr. Davie, gave the results of a comparison of the DNA from a blood sample given by Doheny and that found in semen stains at the crime scene. Two different tests were made. The first test used a multi locus probe in which samples were taken from a number of loci but the precise locations could not be identified. This test produced six matching bands. One was discarded to eliminate overlap with the results of the second test, which will be described later. Using the other five gave a random occurrence ratio of 1 in 840.

The multi locus probe had been superseded by the single locus probe which uses results taken from known locations in the DNA. Three single locus tests were made. Two each gave two matching bands and the third test gave one. Analysis of these five matching bands gave a random occurrence ratio of 1 in 6900. (This ratio would have been even smaller had thirteen locations been used, as is now common.) Davie multiplied the results of both tests to give a ratio of 1 in 5.7 million. Another expert witness, Miss Holmes, had testified that Doheny's blood group was shared by 1 in 7 of the population. Mr. Davie used this too as statistically independent evidence to give, finally, a random occurrence ratio of $(1/840) \times (1/6900) \times (1/7)$ which is 1 in 40.6 million.

Having presented the results of each of his analyses the following exchange took place between Mr. Davie and prosecuting counsel:

Q: What is the combination, taking all those into account?
A: Taking them all into account, I calculated the chance of finding all of those bands and the conventional blood group to be about 1 in 40 million.
Q: The likelihood of it being anyone other than Alan Doheny?
A: Is about 1 in 40 million.
Q: You deal habitually with these things, the jury has to say, of course, on the evidence, whether they are satisfied beyond doubt that it is he. You have done the analysis, are you sure that it is he?
A: Yes.

This sequence of questions and answers was to have great significance. And you can see why.

~ • ~

---

[2]Balding [5] provides much more on DNA profiling.

Five years after the original trial, in July 1996, a second appeal, submitted for Doheny, was heard in the Court of Appeal by Lord Justice Phillips, Mr. Justice Jowitt and Mr. Justice Keen. The grounds for appeal were that "the verdict of the jury was unsafe because the forensic evidence was presented to the jury in a misleading and inaccurate manner and because the substance of that evidence was incorrect."

The judgement, written by Lord Justice Phillips,[3] is of interest because of the way it dealt with statistical aspects of the presentation of DNA evidence. We'll return to that. The decision of the Appeal Court judges rested on three arguments

First, that the cross examination of Mr. Davie given above, and in particular the final question and answer, were clear examples of The Prosecutor's Fallacy. Mr Davie had provided no evidence on which to base his "Yes". The question should not have been put. The presentation of the DNA evidence in this exchange was "inappropriate and potentially misleading".

Second, that the summing up of the evidence in the original trial had been at fault. Evidence had been given by Miss Holmes that a pubic hair found at the crime scene belonged neither to the victim nor to Doheny. In his summing up of the evidence the judge had quite rightly reminded the jury of this. Of Mr. Davie's evidence he said that "...in the end his opinion was that his tests were reliable and they did show that the chances of Mr. Doheny not being the person responsible was so remote as to be possible to discount entirely for all practical purposes." The Appeal judges were critical because it is for the jury, not the judge, to weigh the evidence, even though they were likely to have reached the same conclusion.

Third, that the DNA evidence was faulty. Expert testimony provided by Doheny's counsel at the Appeal raised doubts about Mr. Davie's calculations themselves. The argument was that there was a chance that the two tests, the multi locus and single locus methods, may have identified one or more of the same bands. In this case the results cannot be statistically independent and so the simple multiplication of Mr. Davie would give an incorrect random occurrence ratio. Mr. Davie had seen one such duplication and for this reason reduced from six to five the number of bands used in the analysis of the results of the multi locus probe. The test material had been preserved and so could be re-examined. On the second day of the Appeal counsel for the Crown conceded the point. The effect was to increase the estimate of the random occurrence ratio by a factor of four from 1 in 40 million to 1 in 10 million. Doheny's counsel made other criticisms of the treatment of the multi locus test results including that there was no justification for simply multiplying the different results of those tests.

The Appeal judges concluded that Doheny's conviction "is unsafe and must be quashed".

~ • ~

---

[3]Find the transcript of Lord Justice Phillips' judgement at http://www.bailii.org/ew/cases/EWCA/Crim/1996/728.html. Accessed 26 September 2017.

At the time of the appeal Doheny had already served over half of his sentence and so was eligible for parole. He was released and awarded £125,000 in compensation (some reports say more).

In 1998 he was again jailed for rape. He was released in 2004.

In 2010 he sexually attacked a woman in Manchester city centre. The following year Bolton Crown Court found him guilty and gave an indeterminate sentence for the protection of the public, with a recommendation that he should serve no less than four years.

The quashing of the original sentence was irrelevant to Doheny's subsequent offences—he was due for parole anyway. This case, specifically the appeal, has been much cited because of the arguments used by Lord Justice Phillips and his colleagues in reaching their decision.

~ • ~

Two errors were common to Doheny's trial and that of Sally Clark: the confusion of the probabilities in the cross examination, the first part of The Prosecutor's Fallacy, the transposed conditional, and unjustified assumptions of independence. What is new here is an argument about base rates.

In the Appeal judgement it had been recommended that the result of the multi locus test should have been disregarded. Had this been done the random occurrence ratio would have been the product of the ratios of the single probe tests, 1 in 6900, and the blood group, 1 in 7, which would have given 1 in 48,300. The judgement noted that between experts there were

> ...different approaches to applying the data contained in the population data base used to calculate the results of the tests. Happily we do not have to attempt to resolve this issue, for it is not sufficiently significant to affect the result of this appeal.
>
> The relevant position is this. If one excludes from the calculation the result of the multi locus probe test, in a population base of 800,000 suggested as constituting those within a reasonable proximity of the scene of the crime, there may well have been 20 or so individuals with the same DNA profile as the Appellant and as the crime stain. This figure still renders it an extremely unlikely coincidence that both the Appellant and another of this small cohort should have been in the vicinity of the crime at about the time that it was committed, the more so when one postulates that this other individual would also have had to share with the Appellant the other attributes—general appearance, age and local accent—that the complainant described in her assailant.

Lord Justice Phillips has rounded the ratio of 1 in 48,300 to 1 in 40,000 (0.0025%). In a population of 800,000 twenty individuals, of which Doheny was one, might have been guilty of the assault.

The different approaches to which Phillips refers are made clear by using the full Bayes Grid in Fig. 16.4.

Fig. 16.4 Likelihoods of evidence for Alan Doheny, but no base rates

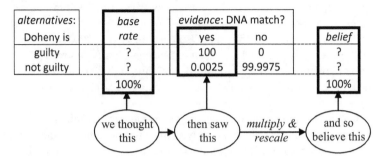

If Doheny was innocent the match probability is 0.0025%. If he was guilty the match would have been certain. But what was it justifiable to think before taking account of the DNA (and blood type) result? What were the appropriate base rates? In the case of Sally Clark there were no other suspects and so on grounds of impartiality there was no initial view (a jury *must* have no initial view) that she was guilty or not. Equal base rates encode this impartiality (Fig. 16.5).

Fig. 16.5 Equal base rates for Sally Clark

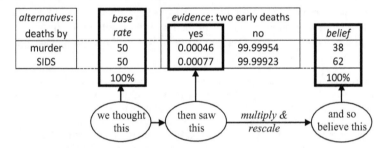

Can this thinking be applied to Doheny? The prosecutor might think so and argue the same impartiality and so equal base rates. The result depends only on the DNA match and so, unsurprisingly, the evidence looks compelling (Fig. 16.6).

Fig. 16.6 Equal base rates for Alan Doheny

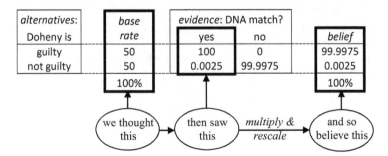

   The defence could argue differently. 800,000 people live close enough to the crime scene that they could have committed the assault. Doheny is one of these so, without any evidence, the chance that he is guilty is 1 in 800,000 or only 0.00013%. This is the correct base rate to use, they will say, it is the only logical place to start. Now, with the DNA evidence as well the probability is just five percent (the 1 in 20 that Phillips gives) that Doheny committed the assault (Fig. 16.7), hardly enough to justify a guilty verdict.

Fig. 16.7 Base rates for Alan Doheny allow for other suspects

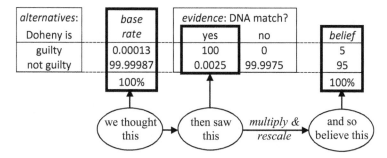

   We are in a tricky but familiar situation. The problem structure which the Bayes Grid has given has made us ask the question but it is still up to us to provide the answer. Once the possibility is raised that others may have been guilty we ask "Yes, but how many?" Doheny's defence estimated the figure to be 800,000, based on proximity to the crime scene. The police could have checked the DNA with the national DNA database to look for a match. This database holds DNA profiles for about one in ten of the UK population. In March 2015 it held profiles for 5,766,369 individuals.[4] Perhaps we should start there, with the size of the database. After all, the attacker may have been passing through and live somewhere else. So why not take the population of the whole country?
   Demonstrating the effect is easy because a convenient assumption can be used to make the methodological point, for example and hypothetically, that the crime happens in a closed island community [6]. But what to do in practice?

---

[4]National DNA Database Strategy Board Annual Report 2014/15. https://www.gov.uk/govern ment/uploads/system/uploads/attachment_data/file/484937/52921_NPCC_National_DNA_Data base_web_pdf.pdf. Accessed 26 September 2017.

Whatever we decide the relevant population to be[5] the number is reduced by other evidence such as a regional accent, age and so on. As the number of potential suspects decreases the probability that Doheny, or any one of the other potential suspects, is guilty increases (Fig. 16.8).

Fig. 16.8 Base rates and suspects

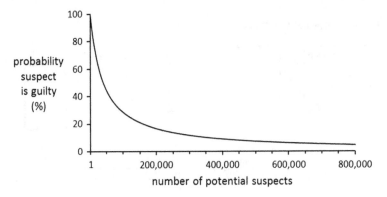

What can the jury do? They will form a view, indeed they must form a view, that is their job. They will have to evaluate each of the pieces of evidence which has been presented to them by lawyers for the defence and prosecution and then combine these somehow to move towards a verdict. Neither is easy.

~ • ~

Lord Justice Phillips talks of other evidence reducing the 800,000. But that other evidence is not free from error. For example, about a third of eyewitness identifications are wrong [8], though not all lead to a conviction.

Reducing the number of potential suspects is not without difficulty.

Most evidence is inherently subject to some uncertainty but for many years this was not thought to be much of a problem. Expert witnesses gave their opinions usually using words to convey the levels of uncertainty in, say, matching shoeprints or blood types. If probabilities were given they were not extreme and so, it was thought, not remote from the experience of jurors. Whether jurors (or any of us) really can cope with interpreting and then combining probabilities is, of course, doubtful as we saw in Chap. 10. But that was the view.

That started to change with DNA evidence and match probabilities of one in several million. It was clear that these probabilities were incomprehensibly small, way beyond everyday experience. Lawyers in court and their expert witnesses had

---

[5]Aitken and Stoney [7] give a brief survey of the issues surrounding sampling in these cases.

carefully to explain to juries what these probabilities meant but this proved to be no easy task, as the appeals of Clark and Doheny showed.

The judicial view of probability could sometimes be hard to fathom as the following two examples show [9]. This is from a case heard in the Family Division of the High Court in England in 1981

> The concept of 'probability' in the legal sense is certainly different from the mathematical concept: indeed it is rare to find a situation in which these two usages co-exist, although when they do, the mathematical probability has to be taken into the assessment of probability in the legal sense and given appropriate weight.

and this from the High Court of Australia in 1984

> If the proposition's validity depends upon logic, and the resolution of a philosophical debate, it seems to me that it cannot properly be characterised as a rule of law. Its validity certainly does not stem from any statutory provision; nor, in my view, can it be argued that its assertion as a proposition of logic in otherwise binding judgments gives it the force of law. If it were shown to be not logically valid, then it could not apply. That would not be the position with a rule of law.

However we interpret these pronouncements they certainly indicate that some judges, and lawyers presumably, made a distinction between reasoning based on numbers which are themselves based on data, and other ways of reasoning to be used with non-numerical evidence. There may also have been an awareness that arguments based on numbers may be privileged over arguments which were not numerical and that this might lead to "trial by statistics". The judge at Sally Clark's trial, Mr. Justice Harrison, put it this way: "...we do not convict people in these courts on statistics. It would be a terrible day if that were so." While this may sometimes be dismissed as a reactionary view we should be careful. Not privileging the quantitative requires fine judgement and an awareness of the danger. This is not easy. But, equally, this does not mean that the quantitative should be disregarded.

Perhaps it would be helpful to express the quantitative weight of evidence in words. The Association of Forensic Science Providers [10] suggests a lexicon (Table 16.1) to show the strength of support provided for a given proposition.[6]

Value of likelihood ratio	Verbal equivalent
1 to 10	weak
10 to 100	moderate
100 to 1,000	moderately strong
1,000 to 10,000	strong
10,000 to 100,000	very strong
more than 100,000	extremely strong

Table 16.1 Lexicon for presentation of evidence

---

[6]Note that categories change as ratios change by a factor of ten. This is just the sort of logarithmic function used in Chap. 9 to measure surprisal

Would this really be an improvement? The expert witness may say that the evidence is "moderate" meaning a likelihood ratio between one in ten and one in a hundred, but what would the juror take it to mean? In addition, we can't do calculations with words: combining evidence with different likelihood ratios expressed in words may not be straightforward. It may well be easier for judges and lawyers and jurors to hear words rather than numbers, but does it help in weighing evidence and reaching a decision? It's hard to see that it does (hard for me, at least).

~ • ~

The principle on which the system of justice rests is that a jury of peers of the accused hears the evidence and then weighs that evidence and finally comes to a decision. How they do all that is left to them; nothing should interfere with their deliberations. The opinion written for the Adams appeal we saw in Chap. 2 makes the point

> ... to introduce Bayes Theorem, or any similar method, into a criminal trial plunges the jury into inappropriate and unnecessary realms of theory and complexity deflecting them from their proper task.

It would be the view of many statisticians that using Bayes' Rule, far from deflecting the jury, will help them to use the evidence to reach a decision in a rational way. But as well as this general objection—that using some mathematical reasoning is at odds with the very idea of a jury trial—there remains the question of whether, and if so how, judgemental assessments might best be used. Here is Lord Justice Phillips again

> In the light of the strong criticism by this court in the 1990s of using Bayes theorem before the jury in cases where there was no reliable statistical evidence, the practice of using a Bayesian approach and likelihood ratios to formulate opinions placed before a jury without that process being disclosed and debated in court is contrary to principles of open justice.

By "reliable statistical evidence" he means data-based numbers such as match probabilities. Before dealing with anything which might be an opinion he wants the method used—Bayes' Rule—to be debated in court so that both prosecution and defence may agree as to the appropriateness of the method.

A clear division is being drawn between the presentation of evidence and the way in which the jury puts it all together to arrive at a decision. Evidence is given by experts who are open to cross examination. A computation using Bayes' Rule (or an alternative) cannot be cross examined, but a proponent of its use can be: a theory cannot be cross examined but a theoretician can. We might think of this person as an expert witness on statistical method in just the same way that a forensic scientist is considered an expert on DNA analysis. Other experts may be called to give evidence from an alternative, perhaps opposing, view. It would be foolish to assert that there is unanimity among statisticians about the use of Bayes' Rule in the way that there is among forensic scientists about evidence based on DNA or blood samples. And so you may agree with the position that probabilities are suitable for presenting evidence [11] but not for reasoning with that evidence.

And yet...

~ • ~

What happens in the jury room is known only to the jurors. A number of studies have simulated jury deliberations by using mock juries made up of students or jury-eligible people. Typically there are two stages. The jury is given a description of the charge and some evidence which does not include probabilities. The jury gives its probability that the accused is guilty. More evidence is then provided which does use probabilities. Because DNA match probabilities are so very small these are usually avoided. Other more comprehensible evidence, such as shoeprints and blood types, is preferred. Having considered this new evidence the mock jury then gives a revised probability of guilt. This is compared with the revised probability found using Bayes' Rule (Fig. 16.9).

Fig. 16.9 Jurors' judgements are conservative[7]

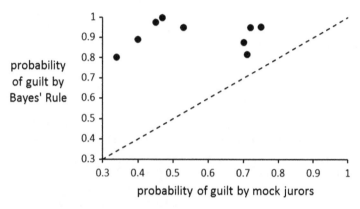

Even though these simulations may be criticised because of their artificiality there does seem to be a consistent pattern: that mock jurors do not update their initial assessments (base rates) as much as Bayes' Rule suggests they should. In other words, they are conservative in their use of evidence.

Interestingly, in some of these simulations using jury eligible subjects those with a college education were more likely to revise their judgements in line with Bayes recommendations [13].

Which is to be preferred, the analysis of Bayes' Rule or the decision making of the human jury? Rationality or accountability?

It is easy enough to make the case for using Bayes as a decision aid. Let the calculations be made according to Bayes rationality, by computer or calculator, and let the jury determine likelihoods and base rates, perhaps changing these values in a sensitivity analysis, and in that way arrive at a decision for which they are prepared to be held to account, even if only by themselves.

This sort of decision aiding happens frequently, in arriving at business decisions, for example. There are cogent objections to its use in the jury room. Most obviously

---

[7]Kaye and Koehler [12], Fig. 1.

that, almost certainly, a facilitator would be needed and this would change the very nature of the deliberation. The concentration on the numerical using methods which may be hard for some jurors to understand may bias outcomes by devaluing the qualitative. Then again, the relations between the different pieces of evidence may involve more than just multiplying base rates and likelihoods. There are ways in which these more complex interactions can be modelled [14, 15] but, useful as they might be, the effect would be to make the job of the jurors even more like that of analysts or clients of analysts.

You may think this to be a wholly desirable development or you may fear that the authority of the jury will have been undermined by the use of a model and by making explicit how their decision is reached. Compromise is likely to be easier using only the ambiguities and imprecision of language.

There seems to be a growing consensus that for forensic and other expert witnesses Bayesian analysis is a useful addition to the methods at their disposal. But as to whether Bayes could or should help that final decision, the jury is still out, as they say.

~ • ~

Weighing all the evidence, some of which is statistical and some not, is not easy and so properly presenting that statistical evidence is important. This is where some Bayes thinking helps.

This is not an issue confined to the courtroom. An unknown number of athletes take banned performance enhancing drugs. Some are caught. In 1988 the sprinter Ben Johnson was stripped of the gold medal he had been awarded for the 100m sprint at the Seoul Olympics.

Lance Armstrong won the Tour de France seven times from 1999 to 2005. In 2012 these titles were removed because of his use of drugs.

The continuing difficulties besetting athletics and other sports are too numerous to list. You will probably recall that Russia was not allowed to participate in the Rio Olympics

And so it goes.

All of which makes the correct presentation of the results of drugs testing as necessary as it is in a court of law. The same difficulties arise, not least The Prosecutor's Fallacy [16, 17]. Confusion over what evidence means is not confined to the courtroom.

~~~ ••• ~~~

References

1. Schneps L, Colmez C (2013) Math on trial: how numbers get used and abused in the courtroom. Basic Books, New York
2. Hill R (2005) Reflections on the cot death cases. Significance 2(1):13–16

3. Joyce H (2002) Beyond reasonable doubt. Plus Mag (21). https://plus.maths.org/content/os/issue21/features/clark/index. Accessed 4 July 2017
4. Thompson WC, Schumann EL (1987) Interpretation of statistical evidence in criminal trials: the prosecutor's fallacy and the defense attorney's fallacy. Law Human Behav 11(3):167–187
5. Balding DJ (2005) Weight-of-evidence for forensic DNA profiles. Wiley, Chichester
6. Dawid AP (1993) The Island problem: coherent use of identification evidence. Department of Statistical Science, University College London, London
7. Aitken CGG, Stoney DA (1991) The use of statistics in forensic science. Ellis Horwood, Chichester, pp 56–57
8. Canter D, Youngs D (2009) Investigative psychology: offender profiling and the analysis of criminal action. Wiley, Chichester, pp 213–214
9. Robertson B, Vignaux GA, Berger CEH (2011) Extending the confusion about Bayes. Mod Law Rev 74(3):444–455
10. Association of Forensic Science Providers (2009) Standards for the formulation of evaluative forensic science expert opinion. Sci Justice 49(3):161–164
11. Aitken C, Roberts P, Jackson G (2010) Communicating and interpreting statistical evidence in the administration of criminal evidence: 1. Fundamentals of probability and statistical evidence in criminal proceedings. Royal Statistical Society, London
12. Kaye DH, Koehler J (1991) Can jurors understand probabilistic evidence. J R Stat Soc Ser A 154(1):75–81.
13. Kaye DH, Hans VP, Dann M, Farley E, Albertson S (2007) Statistics in the jury box: how jurors respond to Mitochondrial DNA match probabilities. J Empir Leg Stud 4(4):797–834
14. Fenton N, Neil M (2013) Risk assessment and decision analysis with Bayesian networks. CRC Press, Boca Raton
15. Kadane JB, Schum DA (1996) A probabilistic analysis of the Sacco and Vanzetti evidence. Wiley, New York
16. Berry DA (2008) The science of doping. Nature 454(7):692–693
17. Faber K, Sjerps M (2009) Anti-doping researchers should conform to certain statistical standards from forensic science. Sci Justice 49(3):214–215

Chapter 17
The Selection Task

The selection task is one of the most frequently used experiments in the psychology of reasoning. A great many subjects, mostly psychology students it seems, have been given handouts like that shown as Fig. 17.1.

Fig. 17.1 The selection task

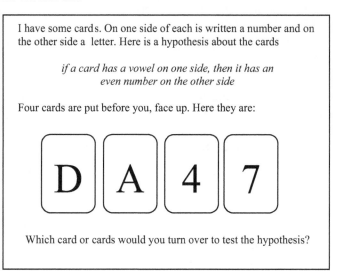

I have some cards. On one side of each is written a number and on the other side a letter. Here is a hypothesis about the cards

if a card has a vowel on one side, then it has an even number on the other side

Four cards are put before you, face up. Here they are:

D A 4 7

Which card or cards would you turn over to test the hypothesis?

Well, what would you do? Before reading on, answer the question.

~ • ~

Bayes' Rule and the Bayes Grid provide a rational way of using evidence to update belief. Likelihoods and base rates are the inputs. Evidence comes from a variety of sources—medical tests, witness statements, DNA analyses, timber cruises and so on. The source of the evidence—which test? which witness? which experiment?—has been taken as given. Someone else decided what to do and Bayes' Rule showed

© Springer International Publishing AG, part of Springer Nature 2018
A. Jessop, *Let the Evidence Speak*, https://doi.org/10.1007/978-3-319-71392-2_17

how to use the evidence obtained. But how were the tests chosen? We'd like to make that selection rationally too.

Often there is no decision to be made because only one test is practically available. But when there are options we must decide: which test to select?

Philosophers and psychologists have for a long time been concerned with this problem of what it means to be rational when making a selection. The usual advice is to use the rules of formal logic, a form of argument dating back to Aristotle. This deals with the relations between propositions and statements which are either true or false. It is unlike Bayesian reasoning which deals with uncertainty measured as probability. Here is an example using two-valued logic.

To reduce the long list of applicants for a job use two criteria, education and experience. Education is indicated by whether or not the candidate has a degree. Experience is indicated by at least five years of relevant work. Think of these as two propositions, that the applicant *has degree* and that the applicant *has experience*. Each proposition is either true or false. We can select in one of two ways. The stricter is to choose only those candidates who satisfy *both* conditions; they have experience AND a degree. If this is thought too strict applicants can be chosen for the next stage in the selection process if they meet *either* criterion; they have experience OR a degree. These two logical conditions are shown in the Table 17.1.

		has experience	
		true	false
has	true	true	false
degree	false	false	false

degree AND experience

		has experience	
		true	false
has	true	true	true
degree	false	true	false

degree OR experience

Table 17.1   Traditional approach

The AND condition is satisfied, is true, only if both *has degree* and *has experience* are true. The OR condition is false only if both are false.

You may have come across this idea as the basis for digital computers where true and false are coded as binary values 1 and 0. Figure 17.2 shows simple function which takes binary inputs and delivers a binary output according to the rule in the box.

Fig. 17.2   Simple rule based processor

If the rule is either AND or OR then tables, just like those above, show how the output is found from the inputs (Table 17.2).

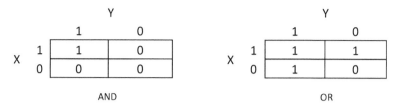

Table 17.2 Binary functions

~ • ~

Simple truth tables can also show the application of *if. . .then* rules. We use these rules all the time

> if it looks like rain then take an umbrella
> if it's too dark to read then switch on the light

and so on. These look quite trivial pieces of common sense, good advice, just follow the rule.

But sometimes we want to test that the rule still holds. Where should we look for useful evidence? The truth table should help by showing what follows from the rule so that we can test if these deductions are true. Take the rule

> if a bird is a swan then it is white

This can be expressed as a deduction (also a prediction) about whether or not a swan is white based on the proposition *all swans are white*

> all swans are white
> this bird is a swan
> therefore this bird is white

This form of making an inference, *this bird is white*, from a proposition, *all swans are white*, is called a syllogism. If the rule is true we are also able to make another deduction which tells us what to conclude if we see a bird which isn't white

> all swans are white
> this bird not white
> therefore this bird is not a swan

What does the truth table look like? The first row comes straight from the rule (Table 17.3) but what of the second row?

		white	
		true	false
swan	true	true	false
	false		

Table 17.3 First row of truth table

It is traditionally thought that other implications are also valid, for instance that *all swans are white* implies that *all non-swans are non-white* and so the truth table should look like Table 17.4.

		white	
		true	false
swan	true	true	false
	false	true	true

Table 17.4 Truth table showing traditional approach

While it is perfectly possible to have something which is not a swan and is white (a snowball) or is not a swan and is not white (a fire engine) it may well seem to you a bit odd that next time you see a fire engine you are at the same time confirming the rule that *all swans are white*. Any time you see anything which is not a swan you are seeing confirmation that all swans are white. This apparently strange state of affairs was pointed out by the philosopher Carl Hempel [1]. To illustrate the difference between results found by applying formal logic and those found by intuition, Hempel used a black raven rather than a white swan, so anything not black and not a raven (that fire engine again) validated the proposition that *all ravens are black*. His puzzle, Hempel's Paradox, is also, more popularly, called the Raven Paradox. Much argumentation followed.

Although it is easy to imagine non-swans they are of no practical value in testing the rule. Sometimes, as we shall see, it may not even be clear that a rule allows us usefully to complete the second row of the table. We can signal this possibility as shown in Table 17.5. We may, as in this case, be able to answer the questions, or we may not.

		white	
		true	false
swan	true	true	false
	false	?	?

Table 17.5 What do we know about non-swans?

In both tables the only way in which the proposition that *all swans are white* can be falsified is to find a non-white swan. You could do this either by noting the colour of swans or looking at non-white things in the hope of finding a swan. Both tactics hold out the possibility of falsification, but you are likely to waste less time by noting the colour of swans, and to be a little surprised if you find one that was not white.

And, as you know already, this is just what happened.

Before Europeans landed in Australia they did indeed believe that all swans were white, but in 1636 the Dutch mariner Antounie Caen saw black swans at Shark Bay, in what is now Western Australia. Some time later, in 1697, another Dutchman,

Willem de Vlamingh, captured two of the birds and returned with them to prove their existence to sceptical Europeans. Nassim Nicholas Taleb, a former derivatives trader and now Professor of Risk Engineering at NYU, used this event for the title of his 2007 book *The Black Swan: The Impact of the Highly Improbable* since when Black Swan has become shorthand for an unforeseen and high impact event.

This form of *if. . .then* argument both gives a rule to follow and also shows where to look for evidence which might falsify the rule. This all looks pretty clear, so what's the problem?

~ • ~

How did you make out with that task at the start of the chapter? The task is repeated here as Fig. 17.3 for convenience.

Fig. 17.3 The selection task

I have some cards. On one side of each is written a number and on the other side a letter. Here is a hypothesis about the cards

if a card has a vowel on one side, then it has an
even number on the other side

Four cards are put before you, face up. Here they are:

D   **A**   **4**   **7**

Which card or cards would you turn over to test the hypothesis?

What did you decide? Using a truth table should help (Table 17.6).

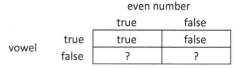

		even number	
		true	false
vowel	true	true	false
	false	?	?

Table 17.6 Partial truth table for the selection task

Table 17.7 shows the partial truth table for this particular selection task shown in Fig. 17.3.

		number	
		4	7
letter	A	true	false
	D	?	?

Table 17.7   shows the truth table for this particular task

It makes sense to turn two cards; the vowel, A, to see if there is an odd number on the reverse side, and the odd number, 7, to see if there is a vowel on the reverse.

This selection task was devised about fifty years ago by Peter Wason of University College [2]. He wanted to investigate the difficulty which many of us have in making this apparently simple selection. The logic of the truth table was taken as the correct recommendation.

There are fifteen possible answers to the question asked in the task: four different single card selections, six pairs, four triples and one selection of all four. The selection recommended by the truth table is one of the pairs, turn cards A and 7. This was not what happened. Fully half of the Wason's subjects chose the cards A and 4. Wason suggested that this was because they thought it seemed to follow directly from the rule; the rule was about nouns and even numbers and so it was thought natural to look at those cards. This is a form of confirmation bias. Very few subjects, about one in six, chose the odd card, 7, either alone or in combination with others. Most people weren't seeking evidence to falsify. Making the same selection as logic recommends was a rarity.

Wason and Diana Shapiro summarised the results of four experiments, 128 student subjects in all [3] (Not all versions of the experiment used the same letters. A, D, 4 and 7 will stand as shorthand for vowels and consonants and odd and even numbers more generally). Table 17.8 shows what they found.

selection	responses	%
A,4	59	46
A	42	33
A,4,7	9	7
A,7	5	4
others	13	10
	128	100

Table 17.8   Results summarised by Wason and Shapiro

These results demonstrate a very strong instinct to verify rather than falsify. It certainly seems that we are not natural logicians. But then we are not often faced with abstract problems such as thinking about letters and numbers written on cards.

Why would we do that? Recognising this, Wason and Shapiro conducted another experiment in which subjects (students again) were allocated to one of two groups. One group had the abstract reasoning task with letters and numbers, as before. The other group were also shown cards but with place names on one side (Manchester, Leeds) and travel modes (car, train) on the other and were asked to test the experimenter's claim that *Every time I go to Manchester I travel by car*. There were sixteen students in each group. As expected, in the first abstract task (letters and numbers) group performance was poor. Only two gave the logically correct solution. In the group with the thematic task (cities and travel modes) this increased to ten correct solutions. The two tasks were formally equivalent but a more realistic framing helped a lot. (We saw something of the framing effect in Chap. 8 with the two forms of question about the serum test for Down's syndrome.)

Perhaps the experimental results of the abstract version of the task show not so much an inability to reason as that unfamiliarity makes reasoning hard. Given a familiar contextual framework subjects may be freed from having to worry about the meaning of cards marked with letters and numbers and so can use more of their cognitive and reasoning capacity to think about what to do. It was easy to envisage what travelling to Manchester meant. Thinking abstractedly just is not something we humans are used to doing, even psychology students.

It is plausible that most of us will be more familiar with the idea of collecting and analysing data than we are with applying syllogistic reasoning. We are quite used to hearing of opinion polls, drugs tests, and market research. Wason and Shapiro thought that although the subjects in their abstract task were told that the rule referred only to the four cards before them they "may, in fact, have regarded the cards as items in a sample from a larger universe, and reasoned about them inductively rather than deductively. In doing this they may have implicitly followed the Bayesian rule which assumes that the probability of a generalization is increased by repetition of confirming instances." We'll return to this thought.

~ • ~

This idea that performance of the selection task improves if it is framed in a more familiar way was taken up by Philip Johnson-Laird, a colleague of Wason's at University College, with Paulo and Maria Legrenzi from the Instituto Superiore di Scienzi Sociale in Trento [4]. At the time of their experiment, 1971, it was the case that both in Britain and in Italy a letter could be posted more cheaply if the envelope was not sealed and so the version of the selection task used asked subjects to imagine that they were postal workers and that their job was to ensure that letters confirmed the following rule: *If a letter is sealed it has a 50 lire stamp on it* (5d in Britain, about 2p in new money). Instead of cards four envelopes were used. The face side showed the address and either a 40 lire or 50 lire stamp. The reverse side showed that the envelope was either sealed or not. This task was attempted only after subjects had given their solutions to the original symbolic letters-and-numbers task. The results were similar to those found by Wason and Shapiro. Of the twenty-four students at University College seventeen gave incorrect solutions to the symbolic task but seventeen gave the correct solution to the realistic mail task. Whether the rule was for British or Italian mail made no significant difference.

It certainly seemed that a realistic framing of the task enabled a better performance. In 1981 Richard Griggs and James Cox of the University of Florida reported the results of their experiments, again with undergraduate subjects [5]. They reported three experiments with similar setups to that used by Johnson-Laird and his colleagues. The first experiment used the cities and transport mode version, for these American students Miami/Atlanta and plane/car. Only three of the thirty-two students gave the solution recommended by logical deduction. Griggs and Cox noted that this was consistent with reports of other experiments which had also failed to show the improvement found by Wason and Shapiro.

The second experiment used the envelope and stamp format, with USA and Mexico as the two postal jurisdictions. Performance was poor. This was plausibly because there was no requirement to pay more for a sealed letter in the United States so that students were unfamiliar with this idea.

The third experiment used a rule which certainly was familiar to students. It was illegal for anyone under nineteen to drink alcohol. The four cards showed beer/coke and over/under nineteen. There was a marked improvement in performance; about three quarters of responses were in accord with syllogistic reasoning.

Griggs and Cox concluded that this was likely to be due to the memory-cueing model which "proposes that performance on the selection task is significantly facilitated when the presentation of the task allows the subject to recall past experience with the content of the problem, the relationship expressed, and a counter-example of the rule . . ."

These results give just a flavour of one of the many strands in a vast literature on the selection task. The question which recurs is just what is being tested in the selection task? It seems to be unarguable that if you apply the recommendation of formal logic you are in a very small minority. It is implausible that we select cards whimsically; something must be happening in our brain. Memory-cueing suggests a test of recollection. Wason and Shapiro's speculation suggests a sampling mindset. Perhaps more is inferred from the rule than was intended, for example that *if vowel then even number* also implies *and if consonant then odd number*. It doesn't, but many of us may think it does.

Is there some other account which describes what we see in the responses to the selection task? Perhaps a little Bayes thinking would help.

~ • ~

Taking a Bayesian view means taking a quite different approach than syllogistic reasoning. This sees the task as making deductions justified by the rule and then looking for falsifying evidence (an idea famously proposed by Karl Popper). The Bayesian approach uses evidence to help in deciding between different models of the relation between letters and numbers. The plural, models, is important. Belief is relative: how much more do I believe this explanation rather than that? If there is only one possible explanation, how can it be rejected? In favour of what? "Probably not that one" sounds a bit weak.[1]

We need likelihoods and base rates. For the selection task there are no data to be modelled by a likelihood distribution. All probabilities must be personal, descriptions of what is believed to be the case. In the majority of experiments subjects do not actually turn the cards but rather are asked what they *would* do. The task resembles the planning stage of an investigation rather than the data analysis which follows.

The work of Tversky and Kahneman, and many others, means that we should be cautious about these personal probabilities. It will be helpful to find how decisions (which cards to turn) are affected by different probability assessments.

~ • ~

In 1994 Mike Oaksford and Nick Chater, then of the Universities of Wales and Edinburgh, published a paper showing how using a Bayesian approach could give an account of the results of selection task experiments [7].

They took the view that instead of a truth table with binary values of either *true* or *false* it was more realistic to think that in tackling the selection problem subjects would consider how likely were the different possible results of turning each card so that instead of thinking

> if this is a vowel there must be an even number on the other side

subjects would think

> if this is a vowel what is the probability of an even number on the other side?

The rule says that there is certain to be an even number. This is interpreted as a probability one hundred percent. Instead of a truth table we have likelihoods, as in the Bayes Grid (Table 17.9).

		number		
		even, 4	odd, 7	
letter	vowel, A	100	0	100%
	consonant, D	?	?	100%

Table 17.9 Likelihood distribution for A card

The benefit of this probabilistic interpretation is that it is easier to think about what happens if the consonant, D, is turned. Although the rule says nothing about

---

[1]It certainly is a bit weak but also very widespread. Saying that a finding is *statistically significant*, which you will have heard often, is based on just this reasoning. Strange but true, despite well rehearsed objections among statisticians and other professionals [6].

this it is clear from experimental results that most people (we) do formulate a view about what might be on the other side of the D card. For example, if we think there is a forty percent chance that there is an even number on the back of the D card the likelihoods are as shown in Table 17.10.

		number		
		even, 4	odd, 7	
letter	vowel, A	100	0	100%
	consonant, D	40	60	100%

Table 17.10 Likelihood distributions for both cards

Using probabilities rather than binary true/false values allows flexibility of description. It could be that the different people will have different probability assessments; twenty percent or thirty percent rather than forty percent. Oaksford and Chater called this the Dependence model.

What alternatives are there to the Dependence model? The simplest is to assume statistical independence. The probability of an even number is the same, forty percent, for both the A and D cards; that is what independence means (Table 11.4). Table 17.11 shows the two models.

	4	7	
A	100	0	100
D	40	60	100

Dependence

	4	7	
A	40	60	100
D	40	60	100

Independence

Table 17.11 Likelihood distributions for two models

There are an infinite number of alternative models between Dependence and Independence which could be described by altering the Independence model so that the probability that there is a 4 on the other side of the A card is something other than forty percent. Oaksford and Chater used just the Dependence and Independence models. In this setup the selection task becomes one of deciding degrees of belief in which of the two models is the more likely to be true. This is a significant shift from interpreting the task as one of syllogistic deduction to choosing a description of the relation between types of letter and types of number.

To do this needs the probabilities of all four results, the four values in the table. To find them needs another assumption: how prevalent are vowels and consonants? It might be thought that since there are only five vowels in the alphabet the probability of a vowel is five out of twenty-five, about twenty percent. This assumes some sort of sampling to select the four cards, as Wason and Shapiro thought might be plausible, even though the task said nothing of this. Alternatively, admitting ignorance argues for equal probabilities, fifty percent for vowel and fifty percent for consonant. Suppose, by whatever reasoning, a value of thirty percent is used. Rescaling the rows of the likelihood tables (Table 17.11) gives the 2 × 2 probability distribution of all four combinations of letter and number (Table 17.12).

	4	7	
A	30	0	30
D	28	42	70
	58	42	

Dependence

	4	7	
A	12	18	30
D	28	42	70
	40	60	

Independence

Table 17.12 Probability distributions for all four combinations for each model

Turning the 4 card would reveal either a vowel, A, or a consonant, D. How likely each result is thought to be depends on which model is used. If the Dependence model is assumed, the probabilities of A and D appearing are 30/58 and 28/58, the left column of the Dependence table in Table 17.12. If the Independence model is assumed the probabilities are 12/40 and 28/40. To get likelihoods for the Bayes Grid rescale each pair to sum to a hundred percent (Fig. 17.4).

Fig. 17.4 Likelihood distributions for both models and evidence from turning the 4 card

alternatives:	evidence: turn 4				evidence: turn 4		
	A	D			A	D	
Dependence	30	28	58		52	48	100%
Independence	12	28	40	→	30	70	100%

Plausibly, there is no strong reason to believe that the Dependence model is more or less likely to be the true description than the Independence model. As we have seen in earlier chapters, this means equal base rates. The Bayes Grid makes the analysis (Fig. 17.5).

Fig. 17.5 Bayes Grid analysis for evidence from 4 card

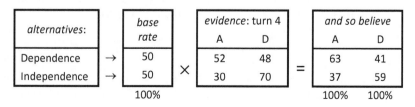

alternatives:		base rate	evidence: turn 4			and so believe	
			A	D		A	D
Dependence	→	50	52	48		63	41
Independence	→	50	30	70	=	37	59
		100%				100%	100%

If the 4 card is turned and shows an A on the reverse side we are justified in believing that the Dependence model is the more likely, probability sixty-three percent. If the reverse shows D then the Independence model is the more likely, probability fifty-nine percent.

Making the same analysis for the other three cards gives the results in Table 17.13.

action	turn A		turn 7		turn 4		turn D	
result	4	7	A	D	A	D	4	7
Dep	71	0	0	59	63	41	50	50
Ind	29	100	100	41	37	59	50	50
entropy (bits)	0.40		0.17		0.04		0	

Table 17.13 Expected result from turning each card

In this Bayesian analysis the purpose of turning a card is to get information from which to learn which of the two models, Dependence and Independence, is the more likely. The entropy measure of information transmission, also called information gain, was described in Table 9.3. It provides an indication of how much might be learned. The last line of Table 17.13 shows the information gain for each of the four cards.

Neither of the two analyses, using deductive logic or using Bayes' Rule, recommend turning the consonant, D. It is not discriminating and so not informative. In the earlier analysis this was because the rule said nothing about this card and so there was no reason to think it relevant. In Oaksford and Chater's analysis there is no information because the likelihoods of finding a 4 or a 7 were the same, forty percent and sixty percent, in both the Dependence and Independence models (Table 17.11). The result is the same, no useful information is to be gained.

The difference between the two approaches is that using the Bayesian analysis all three of the other cards are informative to some extent; the vowel and odd number, A and 7, as before, but now also the even number, 4. This Bayesian information-based approach gives an explanation of why the 4 card might be selected and also of the relative value of each card. This means that the cards can be ranked, A-7-4, according to how informative they are expected to be. Individual subjects may have different threshold values. For instance, some would think it worth turning the 4 card while for others it is insufficiently informative to be useful.

~ • ~

By recasting the selection task as a problem of Bayesian information gathering Oaksford and Chater were able to account for the possible selection of three of the four cards and to give a measure of their relative value. To do this they used two models, Dependence and Independence, as alternative accounts of how subjects might interpret the selection task. The form of each model is fixed but the probabilities are not. Three probabilities are needed

the probability of an even number on the other side of a consonant (40%)
the probability of a vowel card being used for the task (30%)
the base rate probabilities for each model (50%)

This is a much richer framework than the rather austere application of deductive logic. With two alternative models and three parameters (the three probabilities) it is not surprising that a greater proportion of responses can be described. This accounting for results is not necessarily the same as describing a thought process, although the model may suggest a provocative analogy. The differences between what people do in experiments and what the analyses describe is diagnostically helpful in thinking about human reasoning.

Models with more parameters are likely to account for a greater proportion of experimental results. This allows for useful interpretation of the parameter values which may, in turn, lead to a better understanding of behaviour.

Oaksford and Chater needed to see if there were parameter values which gave results corresponding to more of the experimental data than had previous attempts.

There was no reason to believe that unequal base rates were justifiable. They were taken as 50% for each model.

Rather than trying to fix values for the other two probabilities Oaksford and Chater looked for combinations of values that gave particular orderings according to information gain. In the illustration this ordering was A-7-4 (Table 17.13). Card D showed no information gain.

For the competing models row sums were set by choosing the probability that a vowel card was used for the task, thirty percent in this case. These row sums are the same for both the Dependence and Independence models (Table 17.12).

The probability of there being a card with an even number is the sum of the first column of each table. For the Independence model this is, by definition of independence, the probability of an even number on the other side of a consonant which was set at forty percent This is always the case whatever the row sums. But for the Dependence model, because the probability of the A/7 card is always zero, column sums change as row sums change. In this case the probability of a 4 is fifty-eight percent.

Changing the parameter values will in many cases not alter the ordering of the cards, A-7-4. But for some small values of the parameters the changes in information gain do change the ordering. For example, setting the probability of an even number being on the other side of a consonant at four percent and the probability of a vowel card at fifteen percent gives the expected probability of each outcome shown in Table 17.14.

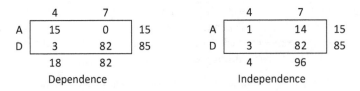

	4	7	
A	15	0	15
D	3	82	85
	18	82	

Dependence

	4	7	
A	1	14	15
D	3	82	85
	4	96	

Independence

Table 17.14 Outcomes for each model with 4% probability of 4 on the other side of D and 15% probability of A

The information gain for each of the three cards is

A   0.88 bits
4   0.35 bits
7   0.08 bits

In this ordering the even numbered card, 4, is more informative than the odd. This was important because it was often what happened. Oaksford and Chater combined the results of 13 studies in which 34 standard abstract versions of the task were reported. There were 835 subjects. The frequency with which each card was chosen was

A   754
4   522
7   215
D   137

The small parameters values which gave this ordering for the first three cards are shown in Fig. 17.6.

Fig. 17.6 Rarity assumption gives ordering A-4-7[2]

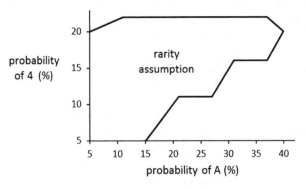

---

[2]Redrawn from [7], Fig. 3, p. 612.

For subsequent analyses Oaksford and Chater restricted parameter values to those in this region, what they called the "rarity assumption". This assumption was justified on two grounds. First, that it was necessary to describe experimental results and so provided an insight on how subjects might interpret the selection task. Second, that subjects do not approach the task without some conception of such tasks and that these frequently involve small probabilities, as in medical diagnosis, for example. Oaksford and Chater's Optimal Data Selection model was not accepted by all [8, 9], yet the increased scope of the account it provides and the useful provocation of the rarity assumption are clear benefits.

In the rest of their paper Oaksford and Chater used this model as a base for looking at other issues associated with the selection task, the non-independence of card selection, for example. Their approach is described in their book *Bayesian Rationality: The probabilistic approach to human reasoning* [10].

~ • ~

This application of Bayes' Rule is unlike others in this book in that its subject is human reasoning and so there is inevitably the thought that if the Bayesian model gives a satisfactory description of what we do might that mean that it also describes how we think? A number of metaphors have been used to describe how the brain might work—as a computer, for example—and certainly Bayesianism is of value at this level. But it is not just that. It also has value as a modelling strategy for problems in cognition and reasoning [11].

Questions about the selection task remain; just why do people select the consonant, D? Alternative explanations continue to be offered [12]. Wason's selection task is the experiment that just keeps giving.

~ • ~

The first three chapters of this section showed applications of Bayes' Rule which used increasingly complex models but the same Bayesian structure as we have throughout this book. The three problems—author attribution, ecology, and radiocarbon dating—were quite technical in nature. The processes which generated the evidence were necessarily described in greater detail than was needed in earlier chapters.

In Chap. 16 the problems were not those of computational complexity but of understanding and using evidence with which, because it was probabilistic, we were unfamiliar and probably uncomfortable. The difficulties here are all too human.

In this chapter the problem was again very human: how to carry out a deceptively simple task. The differences between what we do, what some say we should do, and how to account for those differences was what concerned us. A Bayesian model increased the scope of the analysis.

We have come a long way from Stig and Jan and the Stockholm police but in all cases a little Bayes thinking has helped.

~~~ ••• ~~~

References

1. Hempel CG (1945) Studies in the logic of confirmation (II). Mind 54(214):97–121
2. Wason PC (1968) Reasoning about a rule. Q J Exp Psychol 20(3):273–281
3. Wason PC, Shapiro D (1971) Natural and contrived experience in a reasoning problem. Q J Exp Psychol 23(1):63–71
4. Johnson-Laird PN, Legrenzi P, Legrenzi MS (1972) Reasoning and a sense of Reality. Brit J Psychol 63(3):395–400
5. Griggs RA, Cox JR (1982) The elusive thematic-materials effect in Wason's selection task. Brit J Psychol 73(3):407–420
6. Ziliak ST, McCloskey DN (2008) The cult of statistical significance: how the standard error costs us jobs, justice and lives. The University of Michigan Press, Ann Arbor
7. Oaksford M, Chater N (1994) A rational analysis of the selection task as optimal data selection. Psychol Rev 101(4):608–631
8. Oaksford M, Chater N, Grainger B (1999) Probabilistic effects in data selection. Think Reason 5(3):193–243
9. Oberauer K, Wilhelm O, Diaz RR (1999) Bayesian rationality for the Wason selection task? A test of optimal data selection theory. Think Reason 5(2):115–144
10. Oaksford M, Chater N (2007) Bayesian rationality: the probabilistic approach to human reasoning. Oxford University Press, Oxford
11. Chater N, Oaksford M (eds) (2008) The probabilistic mind: prospects for Bayesian cognitive science. Oxford University Press, Oxford
12. Evans J St BT, Over DE (2004) If. Oxford University Press, Oxford

Chapter 18
Conclusion

Bayes' Rule is the natural way to think about how to evaluate evidence and use it to revise belief.

You are likely to have read this book because you had heard of Bayes but were uncertain what it was. Perhaps you were put off because you thought that it was too mathematical, too difficult. Having got to the end of this book these reservations should have been dispelled. You may now think that Bayes' Rule is just common sense. It is. But, more than that, it is sound statistics too.

To make Bayes' Rule accessible the calculations were shown in a simple table, the Bayes Grid. Making it easy to see the calculation meant that at each application you were reminded (I hope) of the structure of the argument and were provoked to think about the right questions.

~ • ~

If you decide to read more on Bayes' Rule you will almost certainly find the rule written as a formula. You may not recognise it. The note following this conclusion should help.

The range of applications is large and expanding. Software to cope with the necessary computations is becoming available. In Chap. 12 the use of winBugs for MCMC was highlighted.

An important and relatively new use of Bayesian analysis is in helping with complex decision problems. The structure of the problem is represented by a network which shows the interaction between elements of the decision process. This is called Bayesian Network Analysis [1, 2].

~ • ~

Bayes' Rule is a way of learning which imposes a certain discipline.

First, what are the alternative explanations for what we see? The whole point of collecting evidence is to decide how much credibility we should have in the alternatives. The plural is important. If we can think of only one explanation what are we to do but accept it, even with reservations.

© Springer International Publishing AG, part of Springer Nature 2018
A. Jessop, *Let the Evidence Speak*, https://doi.org/10.1007/978-3-319-71392-2_18

Second, do we have a view? The learning process must start from some initial belief. Perhaps we have no strong view. Or perhaps judgement or data mean that we approach the analysis already believing that some alternatives are more likely than others to be the true explanation of the evidence. Base rates encode this starting point.

Third, what is the process by which evidence is generated? We need a model which describes how likely is the evidence given each alternative explanation. This likelihood might be based on an analysis of data (Track Record) or some theoretical model of the process (Margin of Error) or perhaps just some clear thinking (Game Show).

The applications we have seen have been numerical but presented so as to emphasise this structure, for as well as providing a way of calculating it also provides a way of thinking. Sean Carroll, a theoretical physicist at the California Institute of Technology, puts it like this

> Bayes' Theorem is one of those insights that can change the way we go through life. Each of us comes equipped with a rich variety of beliefs, for or against all sorts of propositions. Bayes teaches us (1) never to assign perfect certainty to any such belief; (2) always be prepared to update our credences [beliefs] when new evidence comes along; and (3) how exactly such evidence alters the credences we assign. It's a road map for coming closer to the truth [3].

Carroll is urging that we be careful of assigning base rates of zero because then, however strong the evidence, your (dis)belief will never alter. Pascal, remember, made his argument to convince agnostics, not atheists.

Whether or not you use Bayes' Rule for formal analysis I hope that it will become a habit of your thought: be open-minded and then let the evidence speak.

~~~ ••• ~~~

References

1. Fenton N, Neil M (2013) Risk assessment and decision analysis with Bayesian networks. CRC Press, Boca Raton
2. Jensen FV (1996) An introduction to Bayesian networks. UCL Press, London
3. Carroll S (2016) The big picture: on the origins of life, Meaning and the Universe Itself. Oneworld, London, pp 82–83

A Formula for Bayes' Rule

In this book I have given Bayes' Rule as

belief is proportional to base rate × likelihood

because it was appropriate for you, my not necessarily mathematical reader. If you read more about Bayes' Rule you are likely to find it expressed as a mathematical formula. This note shows how to translate.

A formula is a compact shorthand which saves a lot of typing and, by use of symbols rather than words, shows more clearly what is needed for algebraic manipulation.

Bayes' Rule shows how one probability distribution, belief in alternatives, is found from two others, base rates and likelihoods.

Evidence is used to update initial belief which is expressed as base rates. This initial belief is, by definition of initial, held before or prior to receipt of evidence. Another name for base rates is prior probabilities. The prior probability distribution assigns a probability to each alternative This can be written more compactly by letting P stand for probability and A for alternative to give

base rate = probability of alternative = prior probability = $P(A)$

Next, likelihoods show the probability that different evidence is seen if a particular alternative is true

likelihood = probability of evidence given alternative

Put E for evidence and use the vertical bar | to stand for "given" so that

likelihood = probability of evidence given alternative = $P(E|A)$

© Springer International Publishing AG, part of Springer Nature 2018
A. Jessop, *Let the Evidence Speak*, https://doi.org/10.1007/978-3-319-71392-2

Updating the prior probabilities by multiplying them by likelihoods gives revised probabilities. These describe what we are justified in believing after, or posterior to, receipt of evidence. They are also called posterior probabilities

$$\text{belief} = \text{probability of evidence given alternative}$$
$$= \text{posterior probability} = P(A|E)$$

One last symbol you are likely to see is $\propto$ which stands for "is proportional to". Putting all this together, Bayes' Rule

$$\text{belief is proportional to likelihood} \times \text{base rate}$$

becomes

$$P(A|E) \propto P(E|A) \times P(A)$$

It is common to arrange the expression with likelihood before prior, as shown.

To make this statement of proportionality into a numerical estimate we just rescaled by dividing by the column sum in a Bayes Grid. This column sum was the probability that a particular value of evidence was seen, P(E). The equation form of Bayes' Rule is

$$P(A|E) \ = \ \frac{P(E|A) \times P(A)}{P(E)}$$

You may find this formula written slightly differently—A and B rather than A and E, or prob rather than P—but you'll know what the formula means, however it is written.

～～ ••• ～～

CPSIA information can be obtained
at www.ICGtesting.com
Printed in the USA
LVHW061612170219
607813LV00001B/6/P